# GUIDELINES FOR THE CARE AND USE OF MAMMALS IN NEUROSCIENCE AND BEHAVIORAL RESEARCH

Committee on Guidelines for the Use of Animals in
Neuroscience and Behavioral Research

Institute for Laboratory Animal Research

Division on Earth and Lifes Studies

NATIONAL RESEARCH COUNCIL
*OF THE NATIONAL ACADEMIES*

THE NATIONAL ACADEMIES PRESS
Washington, D.C.
**www.nap.edu**

**THE NATIONAL ACADEMIES PRESS • 500 Fifth Street, N.W. • Washington, DC 20001**

NOTICE: The project that is the subject of this report was approved by the Governing Board of the National Research Council, whose members are drawn from the councils of the National Academy of Sciences, the National Academy of Engineering, and the Institute of Medicine. The members of the committee responsible for the report were chosen for their special competences and with regard for appropriate balance.

This study was supported by Contract/Grant No. N01-OD-4-2139 Task Order 90 between the National Academy of Sciences and the National Institutes of Health. Any opinions, findings, conclusions, or recommendations expressed in this publication are those of the author(s) and do not necessarily reflect the views of the organizations or agencies that provided support for the project. The content of this publication does not necessarily reflect the views or policies of the Department of Health and Human Services, nor does mention of trade names, commercial products, or organizations imply endorsement by the U.S. Government.

**Library of Congress Cataloging-in-Publication Data**

Guidelines for the care and use of mammals in neuroscience and behavioral research / Committee on Guidelines for the Use of Animals in Neuroscience and Behavioral Research.
p. cm.
ISBN 0-309-08903-4 (pbk.) — ISBN 0-309-50587-9 (PDF)
1. Neurosciences—Research—Methodology. 2. Laboratory animals. I. Institute for Laboratory Animal Research (U.S.). Committee on Guildelines for the Use of Animals in Neuroscience and Behavioral Research. II. National Academies Press (U.S.)
RC337.G85 2003
616.8'0427—dc21

2003011672

Additional copies of this report are available from the National Academies Press, 500 Fifth Street, N.W., Lockbox 285, Washington, DC 20001; (800) 624-6242 or (202) 334-3313 (in the Washington metropolitan area); Internet, http://www.nap.edu



Printed in the United States of America

## COMMITTEE ON GUIDELINES FOR THE USE OF ANIMALS IN NEUROSCIENCE AND BEHAVIORAL RESEARCH

**Richard C. Van Sluyters** *(Chair),* University of California, Berkeley, School of Optometry, Berkeley, California
**Michael Ballinger,** Abbott Laboratories, Comparative Medicine, Abbott Park, Illinois
**Kathryn Bayne,** Association for Assessment and Accreditation of Laboratory Animal Care International, Rockville, Maryland
**Christopher Cunningham,** Oregon Health & Science University, Department of Behavioral Neuroscience, Portland, Oregon
**Anne-Dominique Degryse,** Centre de Recherche Pierre Fabre, Laboratory Animal Resources, France
**Ronald Dubner,** University of Maryland Dental School, Department of Oral & Craniofacial Biological Sciences, Baltimore, Maryland
**Hugh Evans,** New York University School of Medicine, Nelson Institute of Environmental Medicine, Tuxedo, New York
**Martha Johnson Gdowski,** University of Rochester, Department of Neurobiology and Anatomy, Rochester, New York
**Robert Knight,** University of California, Berkeley, Department of Psychology, Helen Wills Neuroscience Institute, Berkeley, California
**Joy Mench,** University of California, Davis, Department of Animal Science, Davis, California
**Randy J. Nelson,** Ohio State University, Department of Psychology and Neuroscience, Columbus, Ohio
**Christine Parks,** University of Wisconsin, Madison, Research Animal Research Center, Madison, Wisconsin
**Barry Stein,** Wake Forest University, Department of Neurobiology & Anatomy, Winston-Salem, North Carolina
**Linda Toth,** Southern Illinois University School of Medicine, Division of Laboratory Animal Medicine, Springfield, Illinois
**Stuart Zola,** Yerkes National Primate Research Center, Emory University, Atlanta, Georgia

*Consultants*
**Terrie Cunliffe-Beamer,** Genetics Institute, Andover, Massachusetts
**Peggy Danneman,** The Jackson Laboratory, Bar Harbor, Maine
**Timothy Mandrell,** University of Tennessee, Department of Comparative Medicine, Memphis, Tennessee
**Randall J. Nelson,** University of Tennessee, Department of Anatomy and Neurobiology, Memphis, Tennessee

*Staff*
**Jennifer Obernier,** Study Director
**Kathleen Beil,** Administrative Assistant
**Marsha Barrett,** Project Assistant
**Norman Grossblatt,** Senior Editor

## INSTITUTE FOR LABORATORY ANIMAL RESEARCH COUNCIL

# Preface

Thirteen years ago, a group of 30 some neuroscientists, laboratory-animal veterinarians, and institutional animal care and use committee (IACUC) members gathered at the National Institutes of Health (NIH) in Bethesda, MD, for a workshop sponsored by the National Eye Institute (NEI). The group's purpose was to draft a set of guidelines to help in the preparation and review of protocols for the use of animals in neuroscience. The result was a 45-page report titled *Preparation and Maintenance of Higher Mammals During Neuroscience Experiments*. Published by NIH in 1991, the booklet ultimately went through three printings, and NEI distributed over 30,000 copies to IACUCs, veterinarians, and neuroscientists throughout the world. The Red Book, as it came to be known for its bright red cover, was far more successful than any of the participants in that original workshop ever dreamt it would be.

In the years since the Red Book first appeared, neuroscience research has changed in a number of significant ways. To begin with, it has grown phenomenally. Attendance at the 1989 annual meeting of The Society for Neuroscience was 13,767. By 2001, it had more than doubled to 28,774. Neuroscience institutes, centers, and departments have sprung up in virtually every major research university. The number of students and faculty engaged in neuroscience research is at an all-time high, as is the number of animals that the researchers use in their studies of the nervous system. Like other fields of biomedical research, neuroscience has embraced the use of transgenic methods; as a result, the number of rodents, especially mice, used in biomedical research has increased enormously. With the advent of modern brain-imaging techniques and advances in operant-conditioning methods used to study nonhuman primates, the use of animals in studies of cognitive brain function has risen dramatically. As a result, a number of institutions have formed cognitive-neuroscience centers, where researchers conduct experiments with both human and animal subjects in a two-pronged effort to unravel some of the brain's more complex functions.

Against that backdrop of growth in neuroscience research, the regulatory environment of the use of animals in research has also changed since 1991. The Animal Welfare Act continues to be refined as new policies and regulations are promulgated. In 1996, the National Research Council published the sixth revision of the *Guide for the Care and Use of Laboratory Animals*, a primary source of animal care and use guidance for researchers, veterinarians, and IACUCs. Compliance with the *Guide* is mandated by the Public Health Service as a prerequisite for receiving support from the NIH. The *Guide* is also the standard used by the Association for Assessment and Accreditation of Laboratory Animal Care International to accredit more than 650 animal care and use programs worldwide. The 1996 revision of the *Guide* reiterates the requirement to use professional judgment, as opposed to rigid engineering standards, in the application of many of its recommendations. The extent to which the *Guide* relies on professional judgment has substantially increased, as has the expectation that institutions will develop and use performance-based standards to monitor situations in which professional judgment has an impact on animal welfare.

The task of the Committee on Guidelines for the Use of Animals in Neuroscience and Behavioral Research was to revise and update the guidance provided by the 1991 Red Book and provide information on best use practices for all mammalian species in neuroscience research, not just the "higher mammals" discussed in the 1991 Red Book. Thus, this report presents new information on the use of rodents, including transgenic models. Similarly, whereas the 1991 Red Book offered only a limited discussion of behavioral techniques, this volume includes an extensive coverage of the use of behavioral methods to study brain function. Specifically, the committee was asked to: (1) identify common research themes in contemporary neuroscience and behavioral research based on input from neuroscience and behavioral researchers most familiar with current standards of practice and veterinarian specialists in laboratory animal medicine; (2) exercise collective, professional judgment in applying current animal care and best use practices to procedures in these areas of research; (3) obtain information about new scientific and responsible use developments used to maintain animals during these experiments; (4) prepare a report to serve as an informational resource to assist researchers, laboratory animal medicine veterinarians, and IACUC members in the interpretation and implementation of current standards of practice and promote the training of animal care specialists in this area.

This report, like its predecessor, is intended to provide information and guidance to assist researchers, veterinarians, and IACUCs in interpreting and applying current animal-welfare laws, regulations, policies, and guidelines. It is not meant to replace the *Guide,* nor does it seek to establish policy. It also is not intended to reflect a departure in any way from official animal care and use guidelines.

Like the *Guide*, this report starts with the understanding that a researcher has already decided to use animals in neuroscience research. It is designed to help the neuroscientist prepare an animal-use protocol that provides the information needed by the veterinarian and IACUC that will review it. It is also intended to assist veterinarians and IACUCs in meeting their mandated responsibilities to ensure that complex neuroscience animal-use protocols comply with official animal-welfare regulations and guidelines.

This report has been reviewed in draft form by individuals chosen for their diverse perspectives and technical expertise, in accordance with procedures approved by the National Research Council's Report Review Committee. The purpose of this independent review is to provide candid and critical comments that will assist the institution in making its published report as sound as possible and to ensure that the report meets institutional standards for objectivity, evidence, and responsiveness to the study charge. The review comments and draft manuscript remain confidential to protect the integrity of the deliberative process. We wish to thank the following individuals for their review of this report:

**Michael Fanselow,** University of California, Los Angeles, Department of Psychology, Los Angeles, California
**Roy Henrickson,** Private Consultant, Point Richmond, California
**Julian Hoff,** University of Michigan, Department of Neurosurgery, Ann Arbor, Michigan
**Neil Lipman,** Memorial Sloan-Kettering Cancer Center and Cornell University Medical College, Research Animal Resource Center, New York, New York
**Eric Nestler,** University of Texas, Southwestern, Department of Psychiatry, Dallas, Texas
**Marek Niekrasz,** Northwestern University, Center for Comparative Medicine, Chicago, Illinois
**Gaye Ruble,** Aventis, Inc., Laboratory Animal Science and Welfare Department, Bridgewater, New Jersey
**David Solomon,** University of Pennsylvania, Departments of Neurology and Otolaryngology-Head and Neck Surgery, Philadelphia, Pennsylvania
**Charles Vorhees,** Children's Hospital Research Foundation, Division of Developmental Biology, Cincinnati, Ohio

Although the reviewers listed above have provided many constructive comments and suggestions, they were not asked to endorse the conclusions or recommendations nor did they see the final draft of the report before its release. The review of this report was overseen by:

**Floyd Bloom,** Scripps Research Institute, Department of Neuropharmacology, La Jolla, California
**Marilyn Brown,** Charles River Laboratories, Animal Welfare and Training, East Thetford, Vermont

Appointed by the National Research Council, they were responsible for making certain that an independent examination of this report was carried out in accordance with institutional procedures and that all review comments were carefully considered. Responsibility for the final content of this report rests entirely with the authoring committee and the institution.

No report of this magnitude could be undertaken without the cooperation and expertise of a group of experts, and in this case a special debt of gratitude is owed to the individuals who participated in the workshop that was held, the consultants to the committee, my fellow committee members, and the ILAR staff.

Richard C. Van Sluyters, Chair
Committee on Guidelines for the Use of Animals in Neuroscience and Behavioral Research

# Contents

# Introduction

Like most other biomedical research that uses living animals, experiments in neuroscience and behavior may raise concerns related to the direct physical treatment of animals (for example, surgery, injury, and veterinary care) and concerns about how animals are affected by experimental and general environmental conditions (for example, with respect to distress, well-being, and environmental enrichment). Concerns in the latter category are particularly challenging because defining and assessing such concepts as distress and well-being may not be straightforward.

In an effort to provide information and guidance on the use of mammals in neuroscience and behavioral research, the Committee on Guidelines for the Use of Animals in Neuroscience and Behavioral Research was appointed under the auspices of the Institute for Laboratory Animal Research (ILAR) of The National Academies. The committee was composed of 15 members, both researchers and laboratory animal veterinarians, and were drawn from academia and industry. This committee was funded by the National Institutes of Health and was asked to address the following four items:

1. Identify common research themes in contemporary neuroscience and behavioral research based on input from neuroscience and behavioral researchers most familiar with current standards of practice and veterinarian specialists in laboratory animal medicine.
2. Exercise collective, professional judgment in applying current animal care and best use practices to procedures in these areas of research.
3. Obtain information about new scientific and responsible use developments used to maintain animals during these experiments.

4. Prepare a report to serve as an informational resource to assist researchers, laboratory animal medicine veterinarians, and IACUC members in the interpretation and implementation of current standards of practice and promote the training of animal care specialists in this area.

The ILAR Committee on Guidelines for the Use of Animals in Neuroscience and Behavioral Research (hereafter referred to as the authoring committee) hosted a public workshop on February 27, 2002, to obtain input from leaders in the fields of neuroscience research, behavioral research, and laboratory animal medicine. Following this workshop, the committee met three times during a nine-month period to review the literature, pertinent regulatory documents, and the many references available on the care and use of laboratory animals. After deliberating on responsible use developments and best use practices, the committee drafted this report.

This report provides current information on the care and use of laboratory animals in neuroscience and behavioral research and is aimed at ensuring high-quality, humane care for laboratory animals. Because neuroscience and behavioral research is so diverse, and unique and ambiguous situations continue to arise as science advances, it is impossible for this report to provide specific guidance for every potential research situation. Further, recognizing that every potential research situation cannot be anticipated, there are few regulations or guidelines governing laboratory animal care and use that do not end with the caveat "unless a deviation is justified for scientific reasons and approved by the IACUC." To provide such flexibility in the regulations and guidelines requires the application of professional judgment when applying these regulations and guidelines to each research situation. Often the decisions that must be made are not simple, and reaching effective solutions requires the collective judgment and cooperation of the principal investigator, veterinarian, and IACUC. Therefore, this report emphasizes that developing and evaluating an animal-use protocol requires a decision-making *process,* as many situations do not lend themselves to simple application of regulations and guidelines to reach a yes or no decision.

It is widely held that animal-welfare regulations and guidelines are inflexible and constitute a hindrance to the conduct of high-quality research. One aim of the authoring committee was to highlight the flexibility and promote the use of professional judgment, performance standards, and the decision-making process in evaluating animal protocols, rather than indicating engineering standards. It is the responsibility of institutional animal care and use committees (IACUCs), veterinarians, and researchers to apply creativity and flexibility to balance the needs of high-quality research and humane treatment of animals. In that light, as is the case with other regulatory and guidance documents, the guidelines suggested should not be viewed as laws meant to restrict biomedical research. Rather, they should be interpreted with each unique situation. The guidelines contained in this report are deliberately general. They should be interpreted as a flexible framework that can be applied to diverse research situations to guide decision-making. This publication attempts to demystify the decision-making process that IACUCs, veterinarians, and researchers step through when developing, evaluating, and implementing an animal-research protocol.

This publication is not meant to supersede the guidelines put forth in the 1996 ILAR document *Guide for the Care and Use of Laboratory Animals* (the *Guide*). Instead, it is an informational document that identifies common themes in neuroscience and behavioral research and describes current best practices for animal care and use. It expands on the general guidelines provided in the *Guide*, the Animal Welfare Act, the US Department of Agriculture Animal Welfare Regulations (AWRs) and Policies, the Public Health Service Policy on Humane Care and Use of Laboratory Animals (PHS Policy), and the US Government Principles for the Utilization and Care of Vertebrate Animals Used in Testing, Research, and Training and discusses them as they pertain to the intricacies of neuroscience and behavioral research.

This publication is separated into two main sections. The first section, "General Animal Care and Use Principles and Considerations," contains overarching principles of animal care and use as they pertain to neuroscience and behavioral research. This section includes Chapters 1–3 and includes discussions on the basics of animal husbandry, the definitions of pain and distress, and differentiating between major and minor surgery. Individuals that are well acquainted with the Animal Welfare Act, the Guide, and PHS Policy, may wish to immediately direct their attention to the second section, Chapters 4–9, "Applications to Common Research Themes in Neuroscience and Behavioral Research." This section is based on the six common research themes that the committee identified: survival studies, prolonged nonsurvival studies, studies of neural injury and disease, perinatal studies, agents and treatments, and behavioral studies. This second section discusses current best practices and how they apply to research situations unique to neuroscience and behavioral research.

Chapter 1 covers regulatory and ethical considerations. It identifies the complex and often overlapping regulatory institutions to help neuroscientists to understand the oversight to which they are subject. Chapter 2 deals with the development of animal protocols and issues central to this process, such as euthanasia, minimization of pain and distress, and humane endpoints. The committee emphasizes the important role that the researcher and the veterinarian play in the development of animal protocols and endorses a team approach to developing protocols to prevent misunderstandings. The chapter discusses the use of pilot protocols to evaluate the appropriateness of new animal protocols and underscores that the researcher, veterinarian, and IACUC should collaborate to ensure that the maximal amount of useful preliminary information is collected from these studies. In this chapter, the committee also identifies underused or undervalued methods that researchers and veterinarians can use to assess an animal's well-being, such as monitoring its behavior as a sensitive indicator of its physiologic status. Chapter

3 identifies and discusses the general veterinary and programmatic elements that apply to many types of neuroscience and behavioral research, such as training and supervision of animal handlers, husbandry and nursing care, surgery and procedures, restraint, and food and water regulation. It also discusses the use of genetically modified animals, which have become important neuroscience models in recent years and for which performance-based approaches to care and use are developing constantly. These issues are discussed with emphasis on their application to neuroscience and behavioral research and with emphasis on situations for which the regulations and guidelines are unclear. Throughout Chapters 2 and 3, the text highlights how professional judgment and careful interpretation of the regulations and guidelines contribute to developing performance standards to ensure animal well-being and high-quality research.

Chapters 4 through 9 cover the major experimental themes in neuroscience and behavioral research: survival studies, prolonged nonsurvival studies, studies of neural injury and disease, perinatal studies, studies of agents and treatments, and behavioral studies of neural function. Each chapter highlights the common situations in neuroscience and behavioral research that can pose difficulties for researchers, veterinarians, and IACUCs. Those situations include intended and unintended pain and/or distress, multiple major survival surgeries and modified surgical settings, implantation of devices, and the stresses associated with behavioral paradigms. Recognizing that an experimental protocol can involve elements that are addressed in more than one chapter of this publication, the authoring committee has provided extensive cross-referencing to guide the reader.

Although neuroscience and behavioral research includes widely varied experimental paradigms, each with its own unique animal-welfare concerns, several general animal care and use concerns must be considered in each situation, including:

- Careful monitoring to identify unintended adverse effects.
- Ensuring care for animals that, because of experimental manipulation, may be unable to care for themselves adequately.
- Maintaining an appropriate environment for animals.
- Establishing humane endpoints in advance to avoid or minimize unintended pain and/or distress.

This publication also includes appendix materials that contain information on calculating sample sizes and estimates of the numbers of animals needed to develop and maintain colonies of genetically modified animals. It is difficult for researchers to estimate necessary animal numbers in some situations and the committee included this information to disseminate it to the neuroscience and behavioral research community.

# PART I

## General Animal Care and Use Principles and Considerations

# 1

# Regulatory and Ethical Considerations

The regulatory environment governing the use of animals in neuroscience research is extensive, multilayered, and continuously evolving. Excellent recent reviews of the historical development and current status of that environment can be found in publications by Silverman et al. (2000) and ARENA-OLAW (ARENA-OLAW, 2002). The following is a brief summary of its main elements.

## US ANIMAL WELFARE ACT

The US Animal Welfare Act (AWA; http://www.aphis.usda.gov/ac/awa.html) traces its origins to the Laboratory Animal Welfare Act of 1966. Amended several times over the succeeding years, the AWA names the US Department of Agriculture (USDA) as the federal agency responsible for its implementation and enforcement. Within USDA, the Animal Care unit of the Animal and Plant Health Inspection Service (APHIS/AC) meets this responsibility. Acting under the authority of the AWA, various secretaries of agriculture have developed and promulgated the Animal Welfare Regulations (AWRs; http://www.aphis.usda.gov/ac/publications.html), detailed standards and regulations that govern many aspects of animal care and use programs, including registration of research facilities, institutional animal care and use committees (IACUCs), the attending veterinarian and adequate veterinary care, recordkeeping, reporting, and the procurement, handling, care, treatment, and transportation of animals (9 CFR, Part 2, Subpart C). In addition to the AWRs, there are the Animal Care Policies (APHIS/AC Policies), which were written to further clarify the intent of the AWA. The AWA, AWRs, and the APHIS/AC Policies apply to warm-blooded vertebrates that are bred for use in research—except birds, rats of the genus *Rattus*, mice of the genus *Mus*, and farm animals used in production agriculture. The AWA, AWRs, and the APHIS/AC Policies are available online at http://www.aphis.usda.gov/ac/publications.html.

## PUBLIC HEALTH SERVICE POLICY ON HUMANE CARE AND USE OF LABORATORY ANIMALS

The Public Health Service Policy on Humane Care and Use of Laboratory Animals (PHS Policy) was introduced in 1973 and revised in 1979 and 1986. The PHS Policy (NIH, 1986) applies to all institutions that use live vertebrates in research supported by any component of PHS: the Agency for Health Care Research and Quality, the Centers for Disease Control and Prevention, the Food and Drug Administration, the Health Resources and Services Administration, the Indian Health Service, the National Institutes of Health (NIH), and the Substance Abuse and Mental Health Services Administration. Bolstered by the statutory mandate of the US Health Research Extension Act of 1985 (HREA), the PHS Policy requires institutions to establish and maintain proper measures to ensure the appropriate care and use of animals involved in research, research training, and biologic testing activities. The PHS Policy mandates compliance with the AWA and the AWRs and requires institutions to base their programs of animal care and use on the National Research Council's *Guide for the Care and Use of Laboratory Animals* (NRC, 1996). General administration and coordination of the PHS Policy are the responsibility of the NIH Office of Laboratory Animal Welfare (OLAW). The PHS Policy describes the Animal Welfare Assurance statement, which all covered institutions must submit to OLAW, assuring the office of their compliance with the policy. It also defines the functions of the IACUC, mandates IACUC review of all PHS-conducted or -supported research projects, lists the information required in PHS applications and proposals for awards, and stipulates recordkeeping and reporting requirements. The PHS Policy is available online at http://grants1.nih.gov/grants/olaw/references/phspol.htm.

## GUIDE FOR THE CARE AND USE OF LABORATORY ANIMALS

The National Research Council's *Guide for the Care and Use of Laboratory Animals* (the *Guide*) traces its origin to a 1963 publication by the Animal Care Panel, a group of professionals with an interest in research-animal care that evolved into the American Association for Laboratory Animal Science (AALAS, 2000). The second and all subsequent editions were drafted by committees of the Institute for Laboratory Animal Research and published by the National Research Council. The seventh and most recent edition of the *Guide* was published in 1996 (NRC, 1996). The *Guide* is designed to promote the humane care of animals used in biomedical and behavioral research, teaching and testing; the basic objective is to provide information that will enhance animal well-being, the quality of biomedical research, and the advancement of biologic knowledge that is relevant to humans or animals. It provides guidelines on institutional policies and responsibilities; animal environment, housing, and management; veterinary medical care; and physical plant. In making its recommendations, the *Guide* adopts a performance approach, in which users are charged with achieving well-specified animal-welfare outcomes but can determine individually how best to produce the outcomes, given the constraints and strengths of specific situations. That approach requires that investigators, veterinarians, and IACUCs use professional judgment in designing, reviewing, implementing, and overseeing animal care and use in research, testing, and teaching. Both PHS Policy and the Association for Assessment and Accreditation of Laboratory Animal Care International require that institutions base their programs of animal care and use on the *Guide*. The *Guide* is available online at http://dels.nas.edu/ilar.

## US GOVERNMENT PRINCIPLES FOR THE UTILIZATION AND CARE OF VERTEBRATE ANIMALS USED IN TESTING, RESEARCH, AND TRAINING

The *US Government Principles for the Utilization and Care of Vertebrate Animals Used in Testing, Research, and Training* (US Government Principles) were drafted in 1985 by the Interagency Research Animal Committee (IRAC, 1985), made up of individuals drawn from federal agencies that use or require the use of animals in research or testing. Its nine statements address compliance with the AWA and other applicable federal laws, guidelines, and policies (such as AWRs, HREA, and the *Guide*) and generally provide a set of overarching principles for ensuring that the use of research animals is justified and humane. Compliance with the US Government Principles is mandated by the PHS Policy and the *Guide*. The US Government Principles are available online at http://grants1.nih.gov/grants/olaw/references/phspol.htm.

## ASSOCIATION FOR ASSESSMENT AND ACCREDITATION OF LABORATORY ANIMAL CARE INTERNATIONAL

The Association for Assessment and Accreditation of Laboratory Animal Care International (AAALAC International) is a private, nonprofit organization that promotes the humane treatment of animals in science through a program of voluntary accreditation. Incorporated in 1965, AAALAC International uses the *Guide* as its primary reference document and augments it with reference resources in the peer-reviewed literature. Compliance with AAALAC International's standards is determined through review of an institution's detailed written description of its overall program of animal care and use, which is submitted in advance of a thorough on-site evaluation by a team of AAALAC

International's expert members. Compliant institutions are awarded AAALAC accreditation for a period of 3 years, at the end of which the entire review process is repeated.

## THE 3 Rs

"... by now it is widely recognized that the [most humane] possible treatment of experimental animals, far from being an obstacle, is actually a prerequisite for successful animal experiments."

Russell and Burch, 1959
*The Principles of Humane Experimental Technique*

Every major animal-welfare policy—including the AWA, the PHS Policy, the US Government Principles, and the *Guide*—is based on the principles of the 3 Rs put forth by Russell and Burch in *The Principles of Humane Experimental Technique* (1959). Those principles are:

***Replacement.*** Use of nonanimal systems or less-sentient animal species to partially or fully replace animals.

***Reduction.*** Reduction in the number of animals utilized to the minimum required to obtain scientifically valid data.

***Refinement.*** Use of a method that lessens or eliminates pain and/or distress and therefore enhances animal well-being.

The AWA was amended in 1985, specifically to "reflect the importance of the '3 Rs'" (Hamilton, 1991). The application of the principles was clearly laid out in APHIS/AC Policy 12 "Consideration of Alternatives to Painful/Distressful Procedures":

the regulations state that any proposed animal activity, or significant changes to an ongoing animal activity, must include:

1. a rationale for involving animals, the appropriateness of the species, and the number of animals to be used;
2. a description of procedures or methods designed to assure that discomfort and pain to animals will be limited to that which is unavoidable in the conduct of scientifically valuable research and that analgesic, anesthetic, and tranquilizing drugs will be used where indicated and appropriate to minimize discomfort and pain to animals;
3. a written narrative description of the methods and sources used to consider alternatives to procedures that may cause more than momentary or slight pain or distress to the animals;
4. the written assurance that the activities do not unnecessarily duplicate previous experiments.

The principles of the 3 Rs are also reflected in the *Guide*, which says that the following topics should be considered in the development and review of animal protocols (p. 10):

- Justification of the species and number of animals requested.
- Availability or appropriateness of the use of less-invasive procedures, other species, isolated organ preparation, cell or tissue culture, or computer simulation.
- Appropriate sedation, analgesia, and anesthesia.
- Unnecessary duplication of experiments.

# 2

# Protocol-Development Strategies

## TEAM APPROACH AND SHARED RESPONSIBILITIES

### Building Consensus

A team approach to animal-use protocol development and animal management is valuable for meeting research objectives while maximizing attention to animal care. The team approach relies on the idea of shared responsibilities. Participants include neuroscientists and their laboratory personnel; veterinarians; animal-husbandry staff; the IACUC chair, members, and staff; and the research institution.

The institution must establish a culture of respect for the animals and maintain a commitment to following the *Guide*, PHS Policy, and the AWRs. However, the *Guide,* PHS Policy, and the AWRs are not intended to be barriers to research. Working as a team, the principal investigator (PI), the IACUC, and the veterinarian should be able to devise a means of accomplishing the research goal while addressing the needs of the animals. The "performance standard" encouraged by the *Guide* (p. 3) is a powerful tool for IACUCs and veterinarians to use in developing strategies to promote both animal well-being and good science.

Developing an animal care and use protocol is a negotiation among the PI, the veterinarian, and the IACUC to balance animal well-being with the experimental goals. By involving all parties early in the protocol-development process, particularly for protocols that involve extensive experimental manipulation or difficult to maintain animal models, the PI can help to prevent misunderstandings and delays while facilitating IACUC review and approval.

## Development of Protocols

The PI and research staff develop the scientific concept that underlies an animal protocol. The PI is the expert in the scientific goals and the experimental design and is often highly knowledgeable about animal use and well-being. The veterinarian is the trained expert in the latter subject and can advise the PI and the IACUC on performance-standard development and implementation. However, the IACUC has the final responsibility and authority for evaluating the protocol outcome and approving exceptions to guidelines and regulations if any are necessary. The veterinarian should be involved in animal-use protocol development, preferably before the protocol is submitted to the committee for official review. Indeed, according to the AWRs, the veterinarian must be consulted for any procedure that may cause more than momentary or slight pain or distress (AWR 2.31(d)(1)(iv)(B)). In some cases, veterinary input is gathered primarily during the final protocol-review process, and this delay can complicate protocol review and prolong the approval process.

The IACUC staff and/or chair can solicit additional information and perform a prereview of the animal-use protocol to help the PI recognize portions of the protocol that require clarification or additional information. The IACUC must consider sample size, pain and/or distress, and experimental and humane endpoints among other considerations. The IACUC may request direct observation of a procedure (particularly a new or unusual procedure), pilot studies, and particular experimental measurements or monitoring procedure to evaluate and ensure animal well-being.

## Execution of Protocols

The **research team** has the primary responsibility for animal assessment and intervention. However, researchers and the animal-care staff must coordinate their efforts to provide appropriate animal care and monitoring. Unanticipated adverse effects of the research that are or may be a threat to the health or safety of the animal must be reported to the IACUC immediately. As an animal-use protocol must describe any anticipated adverse effects, if an unanticipated adverse effect that was or could be a threat to the health or safety of the animal were to occur, then the protocol does not accurately reflect the animal-use activity and must be modified accordingly and reapproved.

The **veterinary staff** has the legal responsibility for animal care. Veterinary medical care is best administered with consideration for the scientific goals of the study. However, the veterinarian must have institutional authority to make decisions on behalf of the animal in critical situations.

The **husbandry staff** has day-to-day responsibility for assessment of animal well-being, regardless of experimental use. The caretaker staff is in a unique position to observe large numbers of animals and to understand the animals both individually and as a species or strain. They observe eating, drinking, urination, defecation, behavioral abnormalities, and subtle signs of problems. Because they observe large numbers of normal animals, the animal-care staff can be key participants in determining the phenotype of knockout and transgenic rodents.

The **IACUC**, after review and approval of the animal care and use protocol, has the responsibility to ensure that procedures are carried out in accordance with the protocol. The IACUC may request periodic reports from the PI or from the veterinarian monitoring the research. During semiannual facility inspection, the IACUC can verify that procedures are consistent with the approved protocol.

The **institution** provides the animal-research infrastructure in the form of core facilities for animal care, mandates and resources for training, and the occupational health and safety program. The institutional official (IO) is the individual who is authorized to legally commit on behalf of the research facility that the requirements of 9 CFR parts 1, 2, and 3 will be met (AWR 1.1). The IO is generally the senior administrator with authority to commit institutional resources to ensure compliance with governing regulations and guidelines. The IO should empower the appropriate people to intervene on behalf of the animals, provide adequate facilities and staff, and take the lead in creating a compliant and responsible institutional culture.

## PILOT STUDIES

Pilot studies are integral elements of animal experimentation. As stated in the *Guide* (p. 10), "if little is known regarding a specific procedure, limited pilot studies designed to assess the effects of the procedure on animals, conducted under IACUC oversight, might be appropriate." Numbers of animals in pilot studies are usually small and the researcher and veterinary staff should closely monitor these special kinds of projects. The novelty or unpredictable nature of the experimental techniques or animals being investigated in pilot studies warrants heightened awareness of animal care and welfare.

An important goal for pilot studies should be the collection of the maximal amount of useful preliminary information, and a team effort and team approach will be key factors in reaching the goal. The research and animal-care staff should be aware of potential concerns or complications that may arise during the pilot study. Pilot studies may pose new challenges for all, but the close interactions between scientific and veterinary staff that develop may be the cornerstones for a successful outcome of both the pilot study and the eventual research project.

Examples of situations that may warrant pilot protocols are:

1. The need to develop a new technique.
2. The need to adapt a technique that has not been used previously in a particular species.

3. The need to implement a procedure that uses a technique unfamiliar to the institution except indirectly through published information.
4. The need to modify a procedure to simplify the measurement of a variable or to improve the statistical significance or power of an experiment so as to reduce the number of animals required.
5. The need to refine an existing technique before adapting an approved protocol to use the refinement (testing the results of the refinement may improve the outcome for both the animals and the experimenter.)

## SAMPLE SIZE

As stated in the US Government Principles (IRAC, 1985), investigators should use the minimum number of animals required to obtain valid results (see also the *Guide,* p. 10, and AWR 2.31 (e)(2)). However, investigators frequently err on the side of using too few animals rather than too many (Dell et al., 2002). That results in a study that has too little power to detect a meaningful or biologically significant result. For example, in a meta-analysis of 44 animal experiments involving fluid resuscitation, Roberts and colleagues (2002) found that none had sufficient power to reliably detect a halving of death rate. To avoid this error, researchers should calculate the sample size necessary to detect a statistically significant effect. Several factors must be known or estimated to calculate sample size (Dell et al, 2002):

1. the size of the effect under study (difference between experimental groups)
2. the population standard deviation of the effect
3. the desired power of the experiment to detect the effect (usually 80-90%)
4. the significance level (usually .05 or .01).

Methods for computing sample size are found in Appendix A. In general, the smaller the effect size or the larger the population variability, the larger the sample size must be to detect a difference. It should be noted that using a more sophisticated experimental design and statistical analysis provides more power to detect an effect (Dell et al., 2002).

Some aspects of neuroscience research (such as developing and producing genetically modified animals) pose particular difficulties in estimating the number of animals necessary for a given experiment. Additional guidance on these issues is provided in Chapter 3, "Genetically Modified Animals," and also in Appendix B.

## PAIN AND DISTRESS

Pain may be inherent in the study of pain and/or distress, but it can also be an unintended aspect of the research (for example, in animal models of disease, as a byproduct of a survival surgical procedure, or in transgenic animals with a clinical phenotype). It is critical to recognize and manage animal pain and distress.

The International Association for the Study of Pain has defined pain in humans as an "unpleasant sensory and emotional experience associated with actual or potential tissue damage, or described in terms of such damage" (Mersky, 1979). Although animals cannot communicate verbally, they exhibit motor behaviors and physiologic responses similar to those of humans in response to pain. Those behaviors may include simple withdrawal reflexes; complex, unlearned behaviors such as vocalization and escape; and learned behaviors such as pressing a bar to avoid further exposure to noxious stimulation. However, there are species-specific behaviors that animals may express in response to pain (Bolles, 1970), see Table 2-1 for review.

Stress (or the stress response) has been defined as "the biological response an animal exhibits in an attempt to cope with threats to its homeostasis" (Stokes, 2000). Threats to homeostasis are called "stressors." Stressors can be physical, environmental, or psychologic in origin (NRC, 1992), and adaptation can involve immunologic, metabolic, autonomic, neuroendocrine, and behavioral changes (Moberg and Mench, 2000), but the type, pattern, and extent of the changes depend on the stressor involved. When the animal responds to a stressor in an adaptive way, the animal returns to a state of comfort. It is also possible for stressors to induce responses that have beneficial effects (Breazile, 1987). Animals (and people) are normally exposed regularly to stressors to which they need to respond and adapt (Sapolsky, 1998), and some stress is probably necessary for well-being (NRC, 1992).

When an animal is unable to completely adapt to a stressor and the resulting stress, an aversive state has developed defined as distress. The term distress encompasses the negative psychologic states that are sometimes associated with exposure to stressors, including fear, pain, malaise, anxiety, frustration, depression, and boredom. These can manifest as maladaptive behaviors, such as abnormal feeding or aggression, or pathologic conditions that are not evident in behavior, such as hypertension and immunosuppression (NRC, 1992).

## Regulatory Oversight

Extensive guidelines, policies, and regulations govern the management of pain and distress in laboratory animals. The US Government Principles (IRAC, 1985) state:

> Proper use of animals, including the avoidance or minimization of discomfort, distress, and pain when consistent with sound scientific practices, is imperative. Unless the contrary is established, investigators should consider that procedures that cause pain or distress in human beings may cause pain or distress in other animals [Principle IV].

> Procedures with animals that may cause more than momentary or slight pain or distress should be performed with appropriate sedation, analgesia, or anesthesia. Surgical or other painful procedures should not be performed on unanesthetized animals paralyzed by chemical agents [Principle V].

The AWRs, PHS Policy, and the *Guide* require the IACUC to ensure that animal-use protocols include strategies for minimizing pain and distress in animals. Specifically, USDA (through the AWR 2.31 (d)(ii) and (e) and APHIS/AC Policy 12) requires the investigator to consider alternatives to procedures that

TABLE 2-1 Indicators of Pain in Several Common Laboratory Animals[a]

| Species | General Behavior | Appearance | Other |
|---|---|---|---|
| Rodents | Decreased activity; excessive licking and scratching; self-mutilation; may be unusually aggressive; abnormal locomotion (stumbling, falling); writhing; does not make nest; hiding | Piloerection; rough/stained haircoat; abnormal stance or arched back; porphyrin staining (rats) | Rapid, shallow respiration; decreased food/water consumption; tremors |
| Rabbit | Head pressing; teeth grinding; may become more aggressive; increased vocalizations; excessive licking and scratching; reluctant to locomote | Excessive salivation; hunched posture | Rapid, shallow respiration; decreased food/water consumption |
| Dog | Excessive licking; increased aggression; increased vocalizations, inclusive of whimpering, howling, and growling; excessive licking and scratching; self-mutilation | Stiff body movements; reluctant to move; trembling; guarding | Decreased food/water consumption; increased respiration rate/panting |
| Cat | Hiding; increased vocalizations, inclusive of growling and hissing; excessive licking; increased aggression | Stiff body movements; reluctant to move; haircoat appear rough, ungroomed; hunched posture; irritable tail twitching; flattened ears | Decreased food/water consumption |
| Nonhuman Primate | Increased aggression or depression; self-mutilation; often a dramatic change in routine behavior (e.g., locomotion is decreased); rubbing or picking at painful location | Stiff body movements; reluctant to move; huddled body posture | Decreased food/water consumption |

[a]No single observation is sufficiently reliable to indicate pain; rather several signs, taken in the context of the animal's situation, should be evaluated. The signs of pain may vary with the type of procedure (e.g., orthopedic versus abdominal pain).

may cause more than momentary or slight pain or distress and to provide a written narrative description of the methods and sources used to determine that alternatives to the procedure were not available. As noted in the *Institutional Animal Care and Use Committee Guidebook* (ARENA-OLAW, 2002), the animal-use protocol must provide sufficient information for the IACUC to evaluate the pain and/or distress potentially resulting from the study and the appropriateness of the methods proposed to minimize it. The attending veterinarian is an important team member, working with the researcher and the IACUC to ensure animal welfare when there is the potential for pain and/or distress.

### Assessment of Pain

According to the *Guide,* "fundamental to the relief of pain in animals is the ability to recognize its clinical signs in specific species" (p. 64). Pain can be assessed by evaluating behavioral measures such as eating, socializing, and withdrawal reflexes, and physiologic measures such as heart rate and respiration rate (see Table 2-1). However, species, and even strains and individuals of the same species, may vary widely in their perception of and response to pain (NRC, 1992; Wixon, 1999). Even for an individual animal, pain sensitivity varies among different tissues and organs (Baumans et al., 1994), and pain sensitivities can be altered by pathologic processes or experimental procedures (Carstens and Moberg, 2000). For example, during the initial phase of lipopolysaccharide-induced fever, rats exhibit hyperalgesia, whereas they exhibit hypoalgesia during the later stages of the illness (Carstens and Moberg, 2000). The existence of these differences underscores the point that pain and distress exist as a continuum of experience. In addition, some animals may hide signs of pain; for example, it has been suggested that rats may mask pain during the dark-cycle hours to avoid displaying abnormal activity and increasing their risk of predation (Roughan and Flecknell, 2000).

AALAS (AALAS, 2000) suggests that the magnitude of the pain that the animal is expected to experience be categorized in the protocol and monitored and that there be an opportunity to adjust the pain category once the study is under way. It is important that researchers and animal-care staff have a solid knowledge of the normal and abnormal physiology, behavior, and appearance of the animals in their care (Anil et al., 2002; NRC, 1992).

Acceptable levels of noxious stimulation are those that are well tolerated and do not result in maladaptive behaviors. Acceptable levels range from an animal's pain threshold to its pain tolerance level. Pain threshold is the stimulus level at which pain is first perceived, while pain tolerance is the highest intensity of painful stimulation that an animal will voluntarily accept. As the intensity of a stimulus approaches the pain tolerance level, an animal's behavior will become dominated by attempts to avoid or escape the stimulus, and this degree of pain must be alleviated (Dubner, 1987).

It is important to note that it is usually incorrect to infer that an animal's pain tolerance level is signaled by the onset of avoidance or escape behavior, as some avoidance-escape behavior is an appropriate adaptive response. It is only when the animal's behavior is dominated by avoidance-escape attempts that the behavior becomes maladaptive, signaling unacceptable levels of pain (NRC, 1992).

In pain studies, giving animals control over the source of pain by allowing them to withdraw from a painful stimulus is an effective way to minimize pain and the distress associated with it. If an animal is denied control of the stimulus and it approaches the tolerance limit, maladaptive behaviors will appear, and the animal should be presumed to be in distress. Maladaptive behaviors include persistent attacks on the perceived source of the pain, self-mutilation at the injured or stimulated site, and a state of learned helplessness in which the animal gives up and no longer attempts to escape, avoid, or control the stimulus. To avoid the development of maladaptive behaviors and to minimize pain during experimental manipulations where the animal is denied control of the stimulus, it is critical that the neuroscientist attempt to define the level of pain produced by the stimulus (Dubner, 1987), and ensure that the level imposed by the stimulus is below that which causes the development of maladaptive behaviors. In most cases, previous experimental or published data will indicate the level of pain produced by the stimulus; lacking this information, a pilot study to identify the level of stimulus that produces maladaptive behavior could be useful.

Pain assessment will vary with the pain scale or scoring system used. Scoring systems involve assigning a numeric score to constellations of behavioral, physical, and physiologic observations, and this process can be subjective. There are no generally accepted objective criteria for assessing the degree of pain that an animal is experiencing, and different species or strains can vary in their response to pain. Physiologic measures include heart rate, blood pressure, and respiration rate, but obtaining most of the measures requires some degree of intervention, which may not be feasible or desirable (Baumans et al., 1994).

Recent studies on pain in animals include methods for quantifying specific motor behaviors as indirect measures of responses to mechanical, thermal, or chemical injury (Dubner and Ren, 1999). Animals will withdraw an injured body part from a stimulus, where different levels of stimulation affect the latency or force of withdrawal. This withdrawal response is considered a measure of pain, which correlates highly with more integrative nocifensive behaviors (behaviors in response to pain), such as licking of the injured body part and guarding behavior.

Some behavioral signs are usually associated with pain (Soma, 1987). Animals often communicate through posture. They may exhibit guarding behavior in an attempt to protect the injured part. Vocalizations are important indicators of pain in several species (Anil et al., 2002). Animals in pain may lick, bite, scratch, shake, or rub the site of injury. Restlessness may also be observed, including pacing, lying down and getting up, and shifting weight. Animals in pain may stay in one place for abnormal lengths of time and be reluctant to move or rise. They may withdraw from contact with other animals. They may become listless and refuse to eat or reduce their eating and drinking. They may avoid being handled. These are all possible signs of pain, but none alone is sufficient to determine the presence or level of pain. For example, many animals vocalize intensely when they are handled even if they are not in pain (e.g., Stafleu et al., 1992). Multiple criteria should therefore be assessed (Bayne, 2000; Wallace et al., 1990).

An important step in determining that an animal is in pain is recognition of a departure from normal behavior and appearance (Dubner, 1987; Kitchen et al., 1987; Morton and Griffiths, 1985; NRC, 1992). But as Bayne (2000) indicates, assessments vary with the scale used, and the scales can be very subjective. Flecknell and Silverman (2000) noted that preprocedural scoring is necessary to obtain an appropriate baseline so that confounding variables (such as behavioral effects produced by analgesics) can be identified. For example, some of the consequences of surgery in rats, such as loss of body weight and reduction in food and water intake (signs frequently interpreted to indicate pain or distress), can also be produced in normal, unoperated-on rats by administering opioid analgesics.

Recent evidence indicates that some signs of pain may not be perceived by personnel, such as the ultrasonic vocalizations of infant mice (Nastiti et al., 1991), but are detectable with appropriate equipment. Several excellent references discuss species-specific behaviors that are indicative of pain (Carstens and Moberg, 2000; Hawkins, 2002; Morton and Griffiths, 1985; NRC, 1992; Roughan and Flecknell, in press; Soma, 1987; Wallace et al., 1990).

Assessment of pain should not be influenced by the biases of the observer (Sanford et al., 1986), and the observer should be well trained in both normal and abnormal behaviors of the species in question. Variability among observers can have a substantial effect on the interpretation of assessment data (Holton et al., 1998).

Chronic or persistent pain differs from acute pain because it may not be associated with any obvious pathologic condition and does not serve any protective function. Signs of chronic pain can be subtle and difficult to detect in that an animal's behavior may change slowly and incrementally. Chronic or persistent pain is also more likely to lead to distress and maladaptive behavior. Signs of chronic or persistent pain include decrease in appetite, weight loss, reduction in activity, sleep loss, irritability, and decrease in mating behavior and reproductive performance (Soma, 1987). Alterations in urinary and bowel activities and lack of grooming are often associated with persistent pain. Severe chronic pain can reduce body temperature, cause a weak and shallow pulse, and depress respiration. As noted above, animals cannot control chronic or persistent pain, and it is important to assess the intensity of the pain by using behavioral measures.

## Assessment of Distress

Some of the husbandry and experimental procedures that animals experience in neuroscience and behavioral research have the potential to cause distress. Distress can be either transient or prolonged and can range from mild to severe. Determining when stress becomes distress, and thus an animal-welfare concern that requires amelioration, is difficult. Our understanding of the relationship between the (measurable) physiologic changes that occur during an acute stress response and ensuing adverse psychologic states is generally poor. An animal's mental state can be inferred only indirectly, and many factors can influence whether an animal responds in an adaptive or maladaptive fashion to a particular stressor. Those factors include genetic predisposition, experience, age, sex, species, and the social context in which the stressor occurs.

It is even more difficult to assess the effects of chronic and intermittent stressors that are likely to be experienced by animals as part of routine husbandry and housing in the research setting. There is evidence that animals can successfully adapt physiologically to most stressors experienced as part of routine husbandry and housing (Line et al., 1989; Sharp et al., 2002a,b); however, animals may not be able to physiologically adapt to all such stressors, and their ability to adapt is likely dependent on the aversiveness, duration, and frequency of the stressor (Line et al., 1989). Yet, even when animals can adapt physiologically to such chronic stressors, they may not exhibit behavioral adaptation (Ladewig, 2000). In some cases of chronic stress (for example, in severe depression), the physiologic stress system may eventually stop responding normally to challenges, even after the source of the chronic stress is removed (Sapolsky, 1998).

Nevertheless, it is imperative to evaluate and (where possible) ameliorate distress in the research environment (NRC, 2000). Several general schemes have been proposed for recognizing distress, including a measurable shift in biologic resources (Moberg, 1999), such as a change in metabolic function, or evidence of maladaptive behavior (NRC, 1992). The physiologic and behavioral changes that accompany some states of distress have been fairly well characterized. For example, common manifestations of fear or anxiety are motor tension (shakiness and jumpiness), hyperactivity of the sympathetic nervous system (sweating, increased respiration and heart rate, and frequent urination), and hyperattentiveness (increased vigilance and scanning) (Rowan, 1988).

One problem in assessing stress and distress has been that measurement techniques that involve handling, blood sampling, or tissue collection may themselves be stressors and cause physiologic changes. However, many noninvasive or less invasive methods for physiologic monitoring can now be used in the research setting. They include implanted radio transmitters to measure autonomic nervous function, microdialysis techniques for sample collection, remote blood sampling methods, biosensors for recording central nervous system responses in freely moving animals, and measurement of hormones in hair, feces, and urine (Cook et al., 2000; Goode and Klein, 2002; Koren et al., 2002).

One of the most important noninvasive methods for assessing distress is observation of animal behavior. At least when measured in situations where animals have behavioral flexibility and choice, behaviors can provide information about what animals prefer or avoid, and hence are indicators of emotional states (Mench, 1998). Behavior is a part of an animal's adaptive repertoire for responding to stressors, so it is important to distinguish adaptive behaviors from maladaptive behaviors, such as self-mutilation, unresponsiveness to important signals, hyperactivity, and excessive response to stimulation. Useful observation requires knowledge of the natural history and perceptual capacities of the particular species or strain of animal (Bayne, 1996), the usual frequencies and intensities of particular behaviors, and the causes and functions of the behaviors (Mench, 1998; Rushen, 2000). Because of individual variability, a baseline behavioral profile of an animal should be established if changes in behavior are going to be used to monitor the animal for distress. As with the assessment of pain, personnel assessing behavior should be knowledgeable and skilled in the interpretation of behavior, and assessments should not be influenced by the observers' biases (Bayne, 2000).

## Alleviation of Pain and/or Distress

Four general approaches are available to minimize pain (Dubner, 1987): the use of general anesthesia, the use of local anesthesia and/or analgesia, the training of animals to avoid situations that produce pain (escape-avoidance behavior), and control of the intensity and/or duration of the stimulus by the neuroscientist.

The selection of a general anesthetic should reflect professional judgment as to which anesthetic best meets the clinical and humane requirements without compromising the scientific aspects of the research protocol (NRC, 1996). That sufficient anesthesia has been provided can be ensured by monitoring reflexive responses to painful stimuli, respiration, pupil size, stability of heart rate and blood pressure or electroencephalographic activity. Occasionally in neuroscience research, surgical lesions are created producing a functional decerebration and thereby eliminating all possibility of pain and the need for general anesthesia.

The use of local anesthesia and/or analgesia is also a widely used technique for alleviating pain and/or distress. Vainio et al. (2002) provide a useful description of the clinical efficacy and adverse effects of opioids and nonsteroidal anti-inflammatory drugs (NSAIDS) for analgesia in laboratory animals, as well as the dose, route, and frequency of administration of the common opioids and NSAIDS.

For a discussion of the training of animals to avoid painful stimulation and of how investigators can control the intensity or duration of painful stimulation, see Chapter 4. In some cases, euthanasia may be the most appropriate means of alleviating pain (see Chapter 2—"Humane Endpoints").

Distress is usually an undesirable outcome of an experiment, so strategies for avoiding or minimizing it should be identified during the planning of the study. When distress arises from unintended sources, identification and elimination of the cause is the most obvious course of action. Distress can also be alleviated pharmacologically and nonpharmacologically. Two of the major causes of distress are lack of predictability and control of stimuli (Sapolsky, 1998), and it can be useful to condition animals to experimental or husbandry procedures that they will experience or to allow them to have some control over the stimuli imposed, for example, by providing a way to escape from staged aggressive encounters (as discussed in Chapter 7) or by providing nesting material so that a thermally challenged animal can better control its body temperature. Environmental modifications that can make an animal more comfortable, including changing the ambient temperature, increasing ease of access to food and water, and determining whether social contact would ameliorate or accentuate the stress burden. Appropriate enrichment of the social and structural environment can decrease distress by decreasing boredom and fearfulness, and facilitating coping (Bayne et al., 2002; Carlstead and Shepherdson, 2000; Mench, 1998). If possible, pain-related distress should be managed pharmacologically. However, if pain is the object of the study, pharmacologic options for reducing distress may be few or unavailable. In some studies, the use of sedatives, anxiolytics, dissociative anesthetics, and/or analgesics will not conflict with study goals; IACUCs should require that these options be discussed in animal-use protocols for neuroscience or behavioral research.

In all cases the veterinarian should be consulted regarding methods to minimize pain and distress. Accepted best practices for managing unrelieved pain and distress should be incorporated into the protocol design unless there is a scientific reason to do otherwise. The neuroscientist must provide assurance that unrelieved pain or distress will not continue past the point necessary to achieve the scientific goals of the study (ARENA-OLAW, 2002). Additionally, a mechanism for prompt reporting (for example, to the veterinarian or the IACUC) of animals that have been unexpectedly distressed or pained by the study should be developed and implemented (Bayne, 2000) and should be inherent in the animal-use protocol design.

## USING ANIMAL BEHAVIOR TO MONITOR ANIMAL HEALTH

Animal behavior can be an excellent measure for assessing overall health, indeed, the clinical signs used to diagnose disease in animals are often based on behavior (for example, signs of pain) (Fox, 1968)—although this approach has not been well documented in the veterinary or behavioral literature. A sound understanding of animal behavior is key for the veterinarian or other professional in assessing animal health. Recognition of the importance of behavior as related to animal health, and correspondingly to the veterinary profession, was formalized by the American Veterinary Medical Association in 1993 when the American College of Veterinary Behaviorists (ACVB) was given specialty-board status.

In the research environment, routine behavioral observations can aid in the detection of disease in animals that are not exhibiting any other clinical signs. For example, a cynomolgus monkey was diagnosed with diabetes mellitus through initial observations of uriposia (urine-drinking) (Levanduski et al., 1992); the tentative diagnosis was then confirmed with urinalysis and blood-chemistry evaluation. Sensitive indicators of animal health include measures of food or fluid intake and performance of specific tasks (NIH, 2002).

Because of the long-term use of individual animals in a neuroscience or behavioral study, the physical proximity between researcher and animal, and the wide variety of behavioral data collected during a study, neuroscientists have an excellent opportunity to monitor animal behavior and health. Subtle changes detected in the animal's demeanor or its willingness to work in a study or sudden changes in performance on behavioral tasks may be the first indicators of a health problem that should be investigated. If such changes are noted, the researcher should promptly notify the veterinarian so that the animal can be more fully evaluated.

## HUMANE ENDPOINTS

Endpoints are established for both experimental and humane reasons. An experimental endpoint is chosen to mark the planned end of an experimental manipulation and associated data gathering. A contingent experimental endpoint may also be used to signal euthanasia to remove an animal from the study for humane reasons. On the other hand, in experiments with unrelieved or unanticipated pain/or distress, humane endpoints are criteria that indicate or predict pain, distress, or death and are used as signals to end a study early to avoid or terminate pain and/or distress. Ideal endpoints are those that can be used to end a study before the onset of pain and/or distress, without jeopardizing the study's objectives. However, in most cases, humane endpoints are developed and used to reduce the severity and duration of pain and/or distress (Stokes, 2000).

Humane endpoints should reflect actual or imminent deterioration of an animal's condition, and they should be easy to assess over the course of the study (Toth, 2000). General categories of endpoints include biologic markers, such as the development of paralysis in models of neural tumors (Huang et al., 1993, 1995); markers of therapeutic failure, such as persistent signs of tumor growth despite drug intervention; markers of disability, such as inability to stand in models of bacterial endotoxemia (Krarup et al., 1999); markers of disease exacerbation, such as increased seizure frequency; and general markers of clinical deterioration, such as substantial changes in body weight, alertness, respiration, and body temperature (Redgate et al., 1991; Toth, 1997).

Humane endpoints should be specific to an experimental model or animal strain (Toth, 2000). For example, a decrease in body temperature from 35 to 28°C has been found to be an early predictor of eventual death in studies of bacterial and viral infections, toxicoses, and activity-induced stress in mice (Gordon et al., 1990, 1998; Morrow et al., 1997; Soothill et al., 1992; Toth, 2000; Wong et al., 1997).

The first step in developing a humane endpoint is to describe the clinical progression that a particular animal or group of animals is likely to experience as a result of experimental manipulation or spontaneously occurring disease during their lifetime. Next, potential humane endpoints should be identified. One strategy for identifying humane endpoints is to closely monitor a few animals undergoing a new procedure using score sheets to record clinical, behavioral, and biochemical signs found during the progression of the experiment (Morton, 2000). Finally, a humane endpoint(s) should be selected based on its ability to accurately and reproducibly predict or indicate pain and/or distress, imminent deterioration, or death. The humane endpoint(s) must also be specific to the study (Toth, 2000). For example, the humane endpoint selected for a study of preventative treatments could allow an animal to be euthanized earlier than a humane endpoint selected for a study of disease treatments. In the first instance, the onset of disease symptoms could be a humane endpoint without jeopardizing the scientific goal of studying preventative treatments. However, in the second instance, if the onset of disease symptoms was used as a humane endpoint, the animal would never develop the disease, the treatment could not be tested, and the scientific goal of the study could never be realized.

Humane endpoints may sometimes seem incompatible with experimental endpoints, because ending an experiment for humane reasons can interfere with achieving the scientific goals of the study. The challenge faced by PIs, veterinarians, and IACUCs is to balance the humane treatment of the animal with the scientific goals of the study. Care should be taken when deciding to terminate an experiment early if this will prevent the study from achieving its scientific goals and thereby potentially wastes the animals. It is equally unacceptable to allow animals to experience pain and/or distress beyond the point required to meet the scientific goals of the study (Wallace, 2000).

Humane endpoints should ideally be based on objective criteria and professional judgment (Toth, 1997, 2000) and should be defined in terms that can be understood and recognized by any staff member coming into contact with the animal. For example, the meaning of the phrase "unable to walk" may be more readily understood that the term "moribund" (Krarup et al., 1999). In situations where diseases may occur spontaneously or unexpectedly (for instance with genetically modified animals), the animal-husbandry staff may be the first to identify a subtle change in behavior or appearance that signals a problem and it is important that they understand and can recognize these changes.

Once humane endpoints are established, they should be defined carefully and thoroughly in the animal-use protocol that is submitted to the IACUC for review. The protocol should also establish an adequate but practical frequency of observations and describe the documentation that will be included in an animal's health record. The frequency of observations depends on the nature of the experimental manipulation or disease state and the expected rate of change in an animal's condition.

In some cases, such as genetically modified animals, unpredicted or unintended alterations may occur that adversely affect animal well-being (Stokes, 2000). When developing new types of genetically modified animals, a PI should predict alterations and outcomes based on what is known about the gene of interest, so as to develop humane endpoints (Dennis, 2000). In addition, phenotype screens and measures of general health and well-being may be appropriate to detect unpredicted, adverse alterations in an animal's physiology. If unexpected outcomes do occur, a change in the frequency of observation or an adjustment of the humane endpoint(s) may be warranted.

Another issue is to identify the individuals who will be empowered to decide that a humane endpoint has been reached and that the animal should be removed from the study and/or euthanized. These individuals should be well trained to recognize what is normal and abnormal for the species, and they should clearly understand what is considered an acceptable or unacceptable condition as specified in the animal-use protocol. A clear designation of authority and responsibility to decide on and carry out euthanasia is essential. Ideally, more than one person should have this authority to accommodate for absences. The IACUC should ensure that a designated contact person is listed in the protocol and that someone will be available for consultation or decisions at all times. The responsible veterinarian must have full authority to carry out humane euthanasia when circumstances warrant, although ideally this should be done after consultation with and with the consensus of the research team.

## EUTHANASIA

The *Guide* (p. 10) states that the method of euthanasia should be considered in the preparation and review of animal-use proposals. The AWRs and the *Guide* state that the method of euthanasia must be consistent with the current version of the Report of the AVMA Panel on Euthanasia (AVMA, 2001) unless a deviation is justified for scientific or medical reasons. The AWRs stipulate that guidance on appropriate euthanasia techniques be provided to investigators and animal-care staff by the veterinarian (AWR 2.33 (b)(4)). The AWRs also require that records be maintained on dogs and cats that are euthanized (AWR 2.35 (c)(2)). The IACUC must review and approve the method of euthanasia and must determine whether the proposed endpoint of the study is appropriate, inasmuch as the AWRs further require that "animals that are in severe or chronic pain or distress that cannot be relieved be painlessly euthanized at the end of the procedure or, if appropriate, during the procedure" (AWR 2.31 (d)(1)(v)). The authoring committee notes that requirement does not preclude the development and study of animal models of chronic or persistent pain (AWR 2.31 (d)(1)(iv)(A)); however, animals in severe or intolerable pain should be euthanized. Additionally, animals in studies in which severe pain develops as an unintended consequence should be euthanized or the manipulation causing the unintended pain should be stopped if that would eliminate the pain.

Training staff members to properly perform euthanasia is essential. Training must include instruction both in the specific technique that will be used and in the recognition and confirmation of death (Close et al., 1996). For example, exposure to carbon dioxide can cause deep narcosis that can appear to be, but is not, death. In such cases, animals that superficially appear to be dead may eventually awaken; this arousal can occur after the disposal of carcasses into refrigerators or freezers. The occurrence of death after exposure to carbon dioxide must be confirmed based on careful assessment of the animal for unambiguous signs of death, such as cardiac arrest or fixed, dilated pupils. If an animal is removed from a $CO_2$ chamber before death occurs, the animal either can be returned to the chamber for additional exposure, or, if it is unconscious and nonresponsive, can be humanely euthanized via a physical method (e.g., decapitation or cervical dislocation).

In some species, fear induces animals to become immobile; such immobility must be distinguished from loss of consciousness or death (Close et al., 1996). Some animals release pheromones indicative of fear or distress, which may in turn stress or otherwise disturb other animals (NRC, 1996). Therefore, euthanasia should ideally be performed in an area separate from other animals. However, a recent study suggests that witnessing decapitation may be no more disruptive to Sprague-Dawley rats than other common procedures, such as cage changing, restraint, and injections (Sharp et al., 2003).

Methods of euthanasia that are commonly used in neuroscience research include decapitation, cervical dislocation, carbon dioxide inhalation, and barbiturate overdose. Focused high-intensity microwave irradiation is also used in some cases for measurement of highly labile substances or metabolites (for example, Delaney and Geiger, 1996; Ikarashi et al., 1985; Mayne et al., 1999; Nylander et al., 1997; Theodorsson et al., 1990; Todd et al., 1993). The recommendations of the AVMA Panel on Euthanasia (AVMA, 2001) should be followed unless deviation is justified for scientific or medical reasons (PHS Policy IV(C)(1)(g); APHIS/AC Policy 3; *Guide*, p. 65). However, the AVMA Panel consensus concerning the need for anesthetization prior to decapitation is controversial and is based largely on one publication (Mikeska and Klemm, 1975). Other authors dispute the conclusions drawn from that study, concluding instead that hippocampal and cortical responses to decapitation do not reflect consciousness or resemble the response to painful stimuli (Allred and Berntson, 1986; Vanderwolf et al., 1988), and that instantaneous loss of consciousness, rather than a period of intact pain perception, is likely to occur within a few seconds of decapitation (Bosland, 1995; Derr, 1991; Holson, 1992).

Similarly, there are different views regarding the most humane method for providing euthanasia using carbon dioxide. For example, varying guidance has been provided as to whether it is less distressful to euthanize rodents in a carbon dioxide chamber that has been pre-charged with the gas or not (AVMA, 2001; Close et al., 1996; Hewett et al., 1993; Smith and Harrap, 1997). The most appropriate concentration of carbon dioxide has also been debated, with some authors suggesting that a high concentration promotes a rapid loss of consciousness and death, while others evince that such high concentrations are distressing to the animals (e.g., Danneman et al., 1997). Indeed, recent evidence suggests that carbon dioxide and carbon dioxide-argon mixtures are more aversive to rats and mice than argon alone (Coenen et al., 1995; Leach et al., 2002), although there are a number of reports that carbon dioxide alone provides for a humane death (Hackbarth et al., 2000). Investigators, veterinarians, and IACUCs should be aware of these ongoing debates when they determine the most appropriate method of euthanasia.

The issue of anesthetization or sedation prior to euthanasia is not trivial, because in many circumstances, anesthetizing or sedating an animal before euthanasia, as is recommended for some techniques discussed in the AVMA Panel's report (AVMA, 2001), has adverse consequences in terms of the validity of the experimental design and interpretation of the resultant data. Because anesthetic and sedative agents exert their effects by altering brain function, use of these agents can alter the concentrations, production, or activity of structures or substances that are being examined to answer an unrelated scientific question (for example, Kasten et al., 1990; Mills et al., 1997; Savaki et al., 1980). In such cases, use of anesthesia or sedation may at worst invalidate the study, rendering the animal experimentation useless, but can also cloud the interpretation of the data, perhaps requiring more animals to be tested. Pilot data to confirm this point may be useful in some cases, but under many circumstances, current knowledge about neurophysiologic mechanisms and metabolic regulation may be sufficient to support the conclusion that use of an anesthetic or sedative would confound interpretation of the data. In such cases, it is the responsibility of the investigator to fully describe in the animal-use protocol the scientific evidence that supports any request to withhold anesthetics or sedatives from animals that are to be decapitated (or cervically dislocated), and it is the responsibility of the IACUC to evaluate this evidence carefully to ensure that it provides a compelling rationale for granting an exception to the recommendations of the *Guide* and the AVMA Panel.

## EXPERIMENTAL HAZARDS

Animal use in neuroscience and behavioral research usually does not involve the introduction of physical, chemical, or exogenous biologic hazards. However, any animal use involves the potential for an array of suble physical, chemical, and protocol-related hazards and occasional zoonotic disease risks (NRC, 1997). For example, some research programs involve hazards such as the use of the sodium channel blocker tetrodotoxin or 1-methyl-4-phenyl-1,2,3,6 tetrahydropyridine (MPTP) in basal ganglia research.

Laboratory strains of mice and rats are generally free of infectious agents that pose risks to humans. However, other animals used in neuroscience and behavioral research pose zoonotic risks. Examples of specific risks include those posed by wild mammals (hantavirus, rabies, tularemia, and plague), cats (toxoplasmosis and cat-scratch fever), and nonhuman primates (SIV, B virus, shigella, and tuberculosis) (NRC, 1997).

The key to successful handling of experimental hazards is a systematic process for hazard identification during animal-use protocol development and institutional review. Once hazards are identified, risk management should involve the appropriate safety specialists (NRC, 1997).

One class of hazards associated with neuroscience research that merits special attention is the serious and well-recognized zoonotic diseases associated with nonhuman primates. The most problematic are the viral diseases, notably that caused by the macaque monkey's B virus (also known as Cercopithecine herpesvirus 1). The importance of using awake, behaving rhesus macaques for intensive neurologic study places laboratory personnel at special risk for B virus infection and demands the highest standards of procedural compliance with the use of personal protective equipment, good animal-handling practices, availability of decontaminating equipment, and management of human exposure (Cohen et al., 2002; Holmes et al., 1995; NRC, 2003a). It is essential that laboratories using macaques be well supported by an institutional occupational health and safety program that focuses on the risks of B-virus prevention and control. The basic elements of such a program include procedures and training in dealing with potential exposures, the required use of protective equipment, and access to medical professionals who are knowledgeable about B virus (AAALAC, 2002; CDC-NIH, 1999). Compliance with institutional occupational health and safety requirements should be a prerequisite for IACUC approval of an animal-use protocol and should be evaluated carefully by the IACUC during its semiannual inspections. All macaques, even those from sources thought to be free of B virus and those that repeatedly test serologically negative to B virus, should be presumed to be naturally infected with the virus and handled with appropriate precautions (AAALAC, 2002; NRC, 1997, 2003a).

# 3

# General Animal-Care Concerns

## TRAINING AND SUPERVISION

Oversight and training of all individuals associated with animal care and use (PI, research personnel, students, animal-care staff, veterinary staff, and IACUC members) is critical for the success of research. Gaining consensus on the importance of training is easy; implementation and participation present challenges. Neuroscience research often involves situations in which the research team and the animal-care staff must work in close cooperation to optimize both animal welfare and research outcomes. The diversity in education and experience of these multi-disciplinary teams adds to the training challenge.

Proper training is fundamental in ensuring animal welfare, and is recognized by regulatory agencies. For example, both the AWRs and PHS Policy require institutions to ensure that every person who works with animals is appropriately qualified (AWR 2.32(a) and PHS Policy IV.C.1.f.). There are several good references that provide guidance on training, including *Essentials for Animal Research: A Primer for Research Personnel* (Bennett, Brown, & Schofield, 1994) and *Education and Training in the Care and Use of Laboratory Animals: A Guide for Developing Institutional Programs* (NRC, 1991).

Although the PI is ultimately responsible for ensuring that appropriate training has been provided to the research staff, it is an *institutional* responsibility to make available training in animal anesthesia, surgery, experimental manipulations, and occupational health and safety. The ultimate responsibility for overseeing training rests with the IACUC, which must consider the qualifications of personnel involved in conducting research as part of its protocol review and approval process (AWR 2.31 (d)(1)(viii); NRC, 1996).

Training of research personnel should include procedure-specific training in neuroscience-research techniques, which the PI or senior research staff are usually best suited to teach, and more general training in such subjects as regulation, aseptic technique, anesthesiology, euthanasia techniques, and animal handling, which members of the veterinary or animal-care staff are generally most qualified to teach. The extent of training can depend on the duties and responsibilities of the staff involved. If the procedures to be used have the potential to cause pain and/or distress, mechanisms must be in place to ensure that the research staff can perform them competently. The selection of a trainer should be flexible and adaptive because it will depend on who is best qualified and prepared to provide training. A consortium of individuals from various disciplines may be necessary for complex projects (Kreger, 1995).

Training should be a continuing process. Open communication and cooperation between the veterinary staff and the investigative staff regarding innovations in technique are essential to ensure the most up-to-date and refined use of animals.

Evaluation of outcomes and results is critical in assessing technical experience and the need for training. The IACUC must be prepared to re-review training and experience whenever problems occur in projects.

## MONITORING EXPECTED AND UNEXPECTED CONSEQUENCES

Assessing the nature and context of the clinical problems that an animal may experience during neuroscience experiments can be challenging for both researchers and veterinarians. For example, some strains of genetically modified mice spontaneously develop severe and debilitating disease unrelated to experimental manipulation. In some models, animals may develop substantial or exacerbated neurologic abnormalities because of drug treatment or experimental lesions. The assessment of postprocedure pain, distress, and general health is a matter of subjective clinical judgment that depends on evaluating a variety of measures, including behavioral factors, and recognizing that the interpretation of these measures differs greatly among species; for example, some species mask pain or distress from the observer. However, how a trained animal performs a behavioral task can be a sensitive index of its general condition. Changes in baseline experimental measures can also be informative. In many cases, a change in a specific behavioral measure, rather than changes in a general repertoire of behaviors, is particularly informative. Accordingly, thorough record keeping is essential in any behavioral monitoring program, and the frequency and method of record keeping should be described in detail in the animal-use protocol.

Review of proposed experiments that involve the care and use of animals with induced neurologic deficits poses special concerns for IACUCs. Depending on the nature and extent of the deficits, animals with induced neurologic disease may be limited in their ability to ambulate, obtain food and water, groom, urinate, or defecate, or they may experience pain, behavioral depression, or fear. The experimental induction of debilitating neurologic deficits must be well grounded in scientific need, the animals must receive appropriate specialized care as needed, the number of animals exposed to a debilitating deficit must be minimized, the experimental end point must be well defined (for example, as to the length of time that an animal may be debilitated or the degree of debility), and the experimental protocol should be refined to reduce or eliminate pain, distress, discomfort, and mortality to the greatest extent consistent with valid experimental and statistical design.

Studies of neural injury and disease necessitate stringent requirements for the assessment and alleviation of animal pain and distress. Prolongation or repetition of many treatments, chronic alteration of neural activity, or the destruction of a population of neurons can cause substantial or permanent neurologic deficits. Neuroactive agents and even treatments themselves can cause adverse side effects or toxicity. Evaluating the likelihood of such adverse outcomes and designing strategies for avoiding or alleviating them without compromising the scientific goal of an experiment can be challenging for investigators, veterinarians, and IACUCs. For example, in some studies, the repeated application of an agent or a treatment might require multiple major survival surgeries. In such cases, the stress of undergoing general anesthesia repeatedly, the level of necessary asepsis, and the need to perform the procedures in a surgical setting may be special considerations.

The personnel in a research laboratory usually have some knowledge or expectation about the likely effects of a specific neuroscience procedure on animal health and well-being. Such information is typically solicited as part of the protocol evaluation. A structured approach to developing a profile of anticipated pain, distress, or disease should consider whether any major body systems are likely to be substantially affected during a study. Such an assessment can also guide the development of a systematic approach to animal monitoring and record keeping. The plan should incorporate a list of variables to be assessed and a timetable of observations. Three general considerations apply to research projects that require animal monitoring and maintenance to promote animal well-being (NIH, 1991):

**Consultation.** Consultation between neuroscientists and veterinarians is essential for the design and implementation of monitoring and maintenance procedures. Achieving appropriate solutions to problems that arise in neuroscience experiments requires continuing discussion and collaboration between the PI and veterinarians. Interaction should begin before experiments are initiated. The interaction between the research team and the veterinary staff should provide an opportunity for mutual education and support.

**Responsibility.** Periodic or regular veterinary assessment of both the animal and the experimental records is important in ensuring adequate veterinary care. Animal monitoring and maintenance are conducted and should be documented by neuroscientists as a routine part of their experiments. Documentation should include objective data to identify clinical trends. The records should be readily available to the attending veterinarian acting on behalf of the IACUC. Veterinary oversight is essential to the process for two reasons. First, laboratory-animal veterinarians are trained specialists in the recognition and management of animal health problems, whether spontaneous or iatrogenic. Second, regulatory responsibility for providing appropriate veterinary care rests with the veterinarian.

**Record keeping.** Good record keeping is essential. Records should be written as soon as practical after the animal observation is conducted and should not be phrased with excessive jargon or abbreviations. They should be dated and signed by the record keeper. Observations should be clearly understood by all persons who may have reason to read the records. This documentation serves at least four purposes:

(1) It facilitates detection of gradual changes in health that might not be obvious during a single observation period. A change in condition (such as weight loss) can sometimes be more informative than the condition at any given time.
(2) It requires an advance decision regarding the characteristics that will be assessed and the frequency of monitoring. Completing a form or checklist designed for a particular study promotes diligence and consistency.
(3) It becomes an archive that can be used to improve future study design and animal management.
(4) It documents that appropriate monitoring and maintenance activities were conducted.

All personnel who use animals should be trained to recognize health problems in their animals. That requires knowledge of the appearance and behavior of normal and abnormal animals and a solid understanding of what conditions are acceptable and unacceptable. Animals should be observed initially in an undisturbed state in their home cage. Making such observations can be difficult in some modern high-density caging systems, but attempts should nonetheless be made to evaluate the animals for general activity levels, posture, the condition of the hair coat, signs of self-induced trauma, pattern of respiration, and the general condition of the cage.

Next, the animals should be examined, especially if it is suspected that they have problems. The frequency of individual examination depends on the nature of the debility or disease and the expected rate of progression. For example, for general rodent-colony health surveillance, evaluations should probably be done only on a scheduled basis, such as during a cage change. This limits the number of times that the cages are opened as opening rodent cages properly can be time-consuming and is not risk-free, particularly in light of the possibility of subclinical infectious disease. Characteristics that can be assessed through manipulation are the response to handling; tremors, seizures, vocalization; ulceration; masses; injury; abnormalities of the eyes, ears, nose, or mouth; hyperthermia or hypothermia; and general body condition. Body-condition assessment in rodents requires that personnel learn to palpate the vertebral column to look for emaciation (Ullman-Cullere and Foltz, 1999). Body-condition scoring can be superior to simply weighing animals (especially rodents) because it minimizes the potential for the spread of disease through a shared scale, a reference weight is not needed to calculate a percentage of weight loss for assessment of health status, and body condition can be evaluated more rapidly than body weight.

Neuroscience preparations can cause various degrees of debility that may be predictable in both severity and duration. Sedating the animals at critical post-procedural intervals may prevent discomfort and even inadvertent injury. If debility is unexpectedly severe or prolonged, the PI and the attending veterinarian must intervene to ensure the animal's welfare. For example, a necessary intervention for animals that are not drinking is fluid replacement to prevent dehydration. Similarly, an anorectic animal may be encouraged to eat by being provided easy access to soft, rather than hard, food or a highly palatable food rather than the standard diet.

Appropriate scheduling of procedures that are potentially debilitating, painful, or stressful is important. It may be challenging to provide adequate veterinary care at night, on the weekend, or over holidays because of a shortage of trained staff, closure of diagnostic laboratories, or an inability to obtain specific drugs or equipment. For this reason, it is recommended that researchers schedule experimental procedures that may necessitate supportive or interventional care so that the time during which the animals would be expected to experience distress falls during normal operating hours.

Detection, assessment, and alleviation of pain and distress are additional critical aspects of animal monitoring. Both pharmacologic and nonpharmacologic interventions can be used to alleviate pain and distress. The research and animal-care staff must ensure that instances of animal pain and distress are reported promptly to a veterinarian. Research personnel and animal-care staff must be trained to recognize signs of pain and distress in the species they care for or use.

In summary, appropriate monitoring of animals and maintenance of clinical and experimental records are essential for maximizing the well-being of experimental animals in neuroscience research. Training of personnel and good communication among the research personnel, animal-care staff, and the veterinarian are key components of success.

## ANIMAL HUSBANDRY AND NURSING CARE

Many animal models used in neuroscience research demand exceptional attention to daily care. Induction of neural injury and disease may compromise animals' basic coping and survival mechanisms, as well as their ability to eat, drink, and defecate. Communication, coordination, and creativity in implementation of basic nursing support by the research team and veterinarian are necessary for successful outcomes in these challenging animal models.

The research team should describe the model to the IACUC in the animal-use protocol and should review the approved protocol with the animal-care and veterinary team before beginning the study. The process should include an overview of the scientific benefits that could be achieved from the study and a frank discussion of the challenges involved in maintaining the comfort of the animals after development of a deficit. This front-end investment will go a long way toward creating a team approach to maintaining what are, in effect, intensive-care patients.

The clear delineation of responsibility for monitoring animals is fundamental in ensuring adequate postprocedure care. The *Guide*'s general recommendation for daily observation may be inadequate in many cases. Ideally, frequent observation and the opportunity for intervention constitute a team effort involving both the research group and the animal-care and veterinary staffs. Clearly defined and well-understood scientific goals allow informed intervention (as opposed to inaction) by the caregivers to manage the animals optimally without compromising research goals. A planned strategy for undertaking defined nursing interventions benefits both the animals and the research.

The basics of animal husbandry that are so readily provided in modern housing systems—bedding, food and water, waste-handling—may require extensive modifications or personnel intervention for animals with impaired nervous system function. Enlisting the animal-care group early to consider strategies that will meet basic needs and maximize well-being presents an opportunity to build a team approach.

Generally, recovery from neurosurgery involves the same considerations as recovery from other surgical procedures. Cranial surgery is typically well tolerated by laboratory species. Postoperative analgesia should be used whenever it would not compromise scientific goals. Moistening of chow or providing a diet of softer or more palatable foods for several days postoperatively may make eating more comfortable for the animal and promote food intake, but nutritional modifications are often unnecessary. It may be necessary to consider the use of specialized or modified caging for animals with implanted devices, for example, it may be necessary to remove hanging food bins from rodent cages and place the food on the floor of the cage as hanging food bins could potentially damage a cranial implant.

Special considerations with respect to social housing may be warranted for animals that have had devices implanted for neuroscience research. Animals recovering from such surgery should generally be housed individually during recovery. If damage to implanted devices by a cagemate is unlikely, most animals can then be gradually reintroduced to social housing after their behavior returns to normal.

Some neuroscientists study animal models of human disease. Thus, some surgery is intended to alter the normal physiologic functions of the animal systematically and can affect the psychologic or behavioral state of the animal during postsurgical recovery.

> Such procedures include those that reduce a subject's ability to interact socially or with the environment. Examples are procedures that result in impairment of sensory perception, limit an animal's movement capacities, and impair cognitive abilities. After those procedures, appropriate accommodations should be made in an animal's housing environment or access to enrichment devices to maximize the extent to which it can interact socially and with the environment. Such accommodations can include housing the animal in a social group where it will not be subject to aggressive attacks, giving it manipulanda that can be used with a particular sensory or motor deficit, and giving increased personal attention to an animal that can no longer be put in social housing (NRC, 1998).

On occasion, changes in standard husbandry practices are warranted by the scientific goals of an experiment. For example, cats may be reared in total darkness to determine the influence of visual experience on the development of the visual system (Lein and Shatz, 2000; Mower and Christen, 1985) or animals with lesions of the labyrinth may be housed in the dark to prevent visual compensation for altered vestibular cues (Fetter et al., 1988; Zennou-Azogui et al., 1996). In each of those types of neuroscience research, the animal protocol must ensure appropriate care and monitoring of the animal while maintaining the environmental requirements of the experiment; for example, food and water might be provided in the same locations before and after the lesion is produced.

Care of animals used in neuroscience or behavioral research often requires creativity and exceptions to an institution's normal husbandry procedures. For example, the research team often provides all or much of the daily care of animals used in behavioral studies because of protocol-specific issues or special housing situations. If husbandry responsibilities (including cleaning and sanitization) are to be shared by the animal-care staff and the research staff, the role of each group must be clearly delineated and the care must be documented and freely available to both parties. Integrated husbandry responsibility can work well but only when all members of the team know and accept their roles. The IACUC is authorized to approve exceptions to standard husbandry practices that deviate from the *Guide's* recommendations when the exceptions have a sound justification and appropriate performance standards are met.

## SPECIAL ENVIRONMENTS AND ENCLOSURES AND HOUSING OF MULTIPLE SPECIES

Experimental designs for neuroscience or behavioral studies may involve the use of special environments, including periodic or chronic housing of animals in unusual, nontraditional settings; for example, animals may be reared in total darkness or exposed to omnidirectional sound, microgravity or hypergravity, hyperbaric, or magnetism-free environments. The need to use a special environment may require housing multiple species in close proximity. The *Guide* recommends "physical separation of animals by species to prevent interspecies disease transmission and to eliminate anxiety and possible physiologic and behavioral changes due to interspecies conflict" (p. 58). However, the well-defined health status of most research animals allows the risk of interspecies disease transmission to be reasonably assessed. The possibility of interspecies physiologic and behavioral stressors must also be evaluated. Occasionally, those stressors are an integral part of an experimental design. The veterinarian and IACUC should carefully evaluate such factors and work with the investigator to develop reasonable compromises that allow a balance between animal welfare and research objectives.

Neuroscience or behavioral research may also require the use of nontraditional primary enclosures or caging. Special configurations may allow less space than the standard minimal recommendations in the *Guide*. The *Guide* (p. 25) encourages the use of professional judgment and performance outcomes in assessing space needs for animals with special research needs. It is important that deviations from the *Guide's* space recommendations be evaluated continuously, not just approved prospectively.

## SURGERY AND PROCEDURES

Frequently, surgical procedures are required to meet the scientific needs of neuroscience research, and it is the responsibility of PIs, veterinarians, and IACUCs to ensure that the procedures are designed and conducted in a manner that complies with applicable animal-welfare guidelines and regulations. Interpreting the guidelines and regulations and applying them to a specific neuroscience procedure can be complicated, and it is important for all concerned to be cognizant of the relevant guidelines and regulations.

The *Guide* states that:

> In general, surgical procedures are categorized as major or minor and in the laboratory setting can be further divided into survival and nonsurvival. Major survival surgery penetrates and exposes a body cavity or produces substantial impairment of physical or physiologic functions (such as laparotomy, thoracotomy, craniotomy, joint replacement, and limb amputation). Minor survival surgery does not expose a body cavity and causes little or no physical impairment (such as wound suturing; peripheral-vessel cannulation; such routine farm-animal procedures as castration, dehorning, and repair of prolapses; and most procedures routinely done on an "outpatient" basis in veterinary clinical practice) [pp. 61–62].
>
> Minor procedures are often performed under less-stringent conditions than major procedures but still require aseptic technique and instruments and appropriate anesthesia. Although laparoscopic procedures are often performed on an "outpatient" basis, appropriate aseptic technique is necessary if a body cavity is penetrated [p. 62].

The definition of a major operative procedure in the AWRs is almost identical with that in the *Guide* except that it refers to permanent, rather than substantial, impairment of functions (AWR 1.1). Both the *Guide* and the AWRs offer additional language related specifically to the conduct of survival surgical procedures—those in which the animal is allowed to awaken from surgical anesthesia. The *Guide* provides detailed recommendations regarding facility requirements for survival surgery (pp. 62–63, 78–79) and also states:

> In general, unless an exception is specifically justified as an essential component of the research protocol and approved by the IACUC, nonrodent aseptic surgery should be conducted only in facilities intended for that purpose [p. 62].
>
> The relative susceptibility of rodents to surgical infection has been debated; available data suggest that subclinical infections can cause adverse physiologic and behavioral responses (Beamer, 1972; Bradfield et al., 1992; Cunliffe-Beamer, 1990; Waynforth, 1980, 1987) that can affect both surgical success and research results. Some characteristics of common laboratory-rodent surgery—such as smaller incision sites, fewer personnel in the surgical team, manipulation of multiple animals at one sitting, and briefer procedures—as opposed to surgery in larger species, can make modifications in standard aseptic techniques necessary or desirable (Brown, 1994; Cunliffe-Beamer, 1993). Useful suggestions for dealing with some of the unique challenges of rodent surgery have been published (Cunliffe-Beamer, 1983, 1993) [p. 63].

The AWRs stipulate:

> Activities that involve surgery include appropriate provision for pre-operative and post-operative care of the animals in accordance with established veterinary medical and nursing practices. All survival surgery will be performed using aseptic procedures, including surgical gloves, masks, sterile instruments, and aseptic techniques. Major operative procedures on non-rodents will be conducted only in facilities intended for that purpose which shall be operated and maintained under aseptic conditions. Non-major operative procedures and all surgery on rodents do not require a dedicated facility, but must be performed using aseptic procedures. Operative procedures conducted at field sites need not be performed in dedicated facilities, but must be performed using aseptic procedures [AWR 2.31 (d)(1)(ix)].

When preparing animal-use protocols for neuroscience experiments that require surgical procedures, PIs must take care to describe all aspects of their proposed perioperative procedures accurately and completely. In reviewing the protocols, veterinarians and IACUCs must ensure that proposed surgical procedures are properly classified as major or minor, survival or nonsurvival, and rodent or nonrodent. Furthermore, veterinarians and IACUCs must ensure that the surgical procedures are performed in a manner that complies with the *Guide* and the AWRs; for example, major nonrodent survival surgery should be conducted in a dedicated surgical suite.

Although that sounds relatively straightforward, the complexities of contemporary neuroscience research demand that professional judgment, guided by outcome or performance-based considerations (NRC, 1996, p. 3) be used at each step of the process. Both the PHS Policy and the AWRs permit a great deal of flexibility in their application to research by allowing the IACUC to grant exceptions to their recommendations when acceptable justification is provided. Thus, the PHS Policy states:

> The IACUC shall confirm that the research project will be conducted in accordance with the Animal Welfare Act insofar as it applies to the research project, and that the research project is consistent with the *Guide* unless acceptable justification for a departure is presented [Policy IV.C.1].

The AWRs state:

> In order to approve proposed activities or proposed significant changes in ongoing activities, the IACUC shall conduct a review of those components of the activities related to the care and use of animals and determine that the proposed activities are in accordance with this subchapter unless acceptable justification for a departure is presented in writing; [AWR 2.31 (d)(1)].

A common exception to the AWRs and to the PHS Policy surgical requirements that IACUCs allow is to permit major surgery to be performed in a modified laboratory setting when necessary equipment (such as electrophysiologic recording equipment) cannot be moved to a dedicated surgical suite (see section on "Asepsis and Physical Environment," below).

One area of confusion for IACUCs, veterinarians, and researchers alike is the definition of what actually constitutes a *major surgery*. Neuroscience research often involves procedures that do not meet the strict definitions of major survival surgery given in the *Guide* and AWRs. Some procedures do not involve *both* penetration *and* exposure of a body cavity (for example, endoscopic surgery), or they do *not* penetrate *or* expose a body cavity at all (for example intravenous infusion or injection of neuroactive or neurotoxic substances, closed-head trauma, or peripheral neurectomy). Determining whether such procedures meet the definitions of major survival surgery hinges on whether they seem likely to produce "substantial impairment of physical or physiological functions" (NRC, 1996, pp. 11-12, 61) or "permanent impairment of physical or physiological functions" (AWR 1.1).

The IACUC must assess whether a proposed minimally invasive procedure seems likely to result in an impairment of physical or physiologic functions that is substantial or permanent. If so, the procedure must be categorized as a major surgical procedure and reviewed as such by the IACUC to ensure compliance with the provisions of the *Guide* and the AWRs. However, both the *Guide* and the AWRs expect the IACUC to exercise professional judgment in applying their criteria to the review of surgical protocols. For example, the *Guide* does not define what constitutes a "substantial impairment of physical or physiologic functions," and does not require that an induced impairment be permanent to be considered major surgery. The AWRs stipulate that a noninvasive procedure should result in a "permanent" impairment to be classified as major surgery, but they do not require the impairment to be substantial. Because minimally invasive procedures like those mentioned above can result in impairments whose severity ranges from no apparent loss of function to obvious major functional deficits or impairments whose severity changes markedly (either lessening or worsening) over time, IACUC review of these kinds of protocols can be challenging.

Rather than debate the extent to which a particular neuroscience procedure meets the various regulatory criteria for a major surgical procedure, the authors of this report strongly recommend that PIs, veterinarians, and IACUCs collaborate on the development of animal-use protocols that are designed to safeguard animal welfare and address the scientific needs of the research. The *Guide* and the AWRs provide for sufficient flexibility in the application of their standards for major surgical procedures to allow those involved to craft such protocols. Careful attention should be given to the outcomes of earlier neuroscience studies that used the same or similar procedures. In the absence of precedents for a particular minimally invasive procedure, consideration should be given to obtaining preliminary data from a pilot study performed under direct veterinary supervision and with appropriate reporting to the IACUC.

## Multiple Major Survival Surgeries

In general, multiple major survival surgeries are discouraged, but they may be conducted if they are scientifically justified, related components of a research project approved by the IACUC (NRC, 1996). For example, animals that receive a unilateral visual cortex lesion neglect visual stimuli presented to them on the side opposite the lesion (contralateral visual neglect). Subsequent lesioning of the tectum can ameliorate this neglect (Sprague, 1966). A physical or chemical lesion of the basal ganglia causes a Parkinson's-like tremor in animals that can be reduced or eliminated by a second lesion (Bergman et al., 1990; Wichmann and DeLong, 1996) or by stereotaxic implantation of a stimulating device (Benazzouz et al., 1993, 1996; Boraud et al., 1996). The use of cranial implants for experimental restraint, recording chambers, or implanted monitoring devices is another situation where multiple survival surgeries may be justified.

Some neuroscience-research designs involving multiple surgeries and procedures may have special requirements for asepsis and facilities that will be discussed later. Careful monitoring of the animal, in consultation with the veterinary staff, is necessary. Techniques should be continually refined to minimize pain and/or distress and the monitoring program should be appropriately matched to the anticipated level of pain and/or distress.

Multiple major survival surgeries may also be used to conserve scarce animal resources or if the multiple major surgeries are needed for clinical reasons (NRC, 1996). When a research project involves species covered by the AWA, a waiver must be obtained in writing from USDA for multiple major surgeries that are *not* related components of a research project (APHIS/AC Policy 14). The balance between the welfare of the individual animal and minimization of the numbers of animals used must be carefully weighed by the IACUC. Cost alone is not an acceptable reason for performing multiple major surgeries (NIH, 1991).

## Planning for Survival Procedures

In survival-surgery experiments, monitoring and maintenance issues arise before, during, and after surgery, and in relation to long-term survival and animal health. Responsibility for and details of monitoring of individual animals during and after surgery must be clearly defined in the animal-use protocol (APHIS/AC Policy 3) and presurgical planning should identify the personnel who will perform these duties (NRC, 1996). Medical records should be maintained on each animal throughout the course of an experiment; in fact, this is a requirement for all AWA covered species (APHIS/AC Policy 3). Before any surgery, the weight, general health, and distinctive characteristics of the animal should be noted.

Careful consideration should be given to the anesthetic agents used so that adverse effects of the agents on data collection can be minimized (Cherry and Gambhir, 2001) and the need for post-operative analgesics reduced. Provisions should be made to monitor vital signs and depth of anesthesia frequently, and methods to monitor the animal before and after the surgical procedures should be established.

Removal of food is a standard veterinary practice before any major or minor recovery surgical procedure (Flecknell, 1996), although water should not be restricted. However, as rodents and rabbits cannot vomit, it is unnecessary to fast them prior to a surgical procedure (Waynforth et al., 2003). To maintain proper hydration throughout the surgical procedure, an intravenous line may be established through which supplemental doses of anesthetic or emergency drugs may also be delivered when appropriate. This is especially encouraged for larger animals such as nonhuman primates. Administration of fluids is especially important for smaller mammals during lengthy surgical procedures because their ratios of surface area to body weight and higher metabolic rates necessitate nearly double the fluid supplementation necessary for larger mammals (Balaban and Hampshire, 2001). Fluids should be warmed before infusion to prevent their contributing to hypothermia (Balaban and Hampshire, 2001). Maintaining body temperature during the surgical procedure and post-operative recovery is critical, as a side effect of sedation is hypothermia. Rodents and other small mammals are particularly susceptible to irreversible hypothermia leading to death (Hedenqvist and Hellebrekers, 2003).

If surgery facilities are not in the vivarium, the method and route of transportation to and from the surgery facilities should be considered when preparing animal-use protocols. When these routes take the animal through public areas, such as hospital corridors, care must be taken to minimize any potential contact with the public, who are not expecting nor are prepared to encounter the animal.

Special attention must be given to the public health concerns, such as B virus exposure, that may arise from transporting an animal through public areas. Procedures should be established for dealing with emergencies that may arise during transport, such as bites, scratches, or splashes to a member of the public or the research staff. Consideration must also be given to the potential for animal tissues or fluids to contaminate public corridors, elevators, or patient areas during the transport. Furthermore, the personnel and methods used to monitor the animals and to administer appropriate care to ensure their well-being during transport should be identified. Planning for the transfer of an animal from a vivarium to a surgery facility should include all personnel that will be involved in the transport.

## Anesthesia and Analgesia

The goal of this section is to provide investigators, veterinarians, and IACUCs with a general understanding of the differences between anesthetics and analgesics and the concepts underlying preemptive analgesia and balanced anesthetic regimens. The purpose underlying the use of any of these drugs or regimens is to relieve unintended pain and/or distress (experiments involving unrelieved pain and/or distress are discussed later in this chapter). As noted in the US Government Principles (IRAC, 1985), "proper use of animals, including the avoidance or minimization of discomfort, distress, and pain when consistent with sound scientific practices, is imperative".

As noted in the *Guide* (p. 64),

> The selection of the most appropriate analgesic or anesthetic should reflect professional judgment as to which best meets clinical and humane requirements without compromising the scientific aspects of the research protocol.

The use of professional judgment, open discussion, and the flexibility of all involved parties are particularly encouraged when tackling this complex issue.

In developing a pain-relieving regimen, it is important to understand the difference between anesthesia and analgesia. General anesthesia produces a loss of awareness or consciousness and is used for surgical procedures or experiments that cannot be conducted in awake animals (NRC, 1992). Examples of general anesthetics are inhalation anesthetics, such as isoflurane; opioids, such as fentanyl; and dissociatives, such as ketamine. Inhalation anesthetics produce unconsciousness and muscle relaxation sufficient for surgical intervention (NRC, 1992). However, many injectable anesthetics do not provide enough sedation, muscle relaxation, or analgesia to be used alone. For example, fentanyl provides sedation and analgesia, but muscle relaxation is poor (Hedenqvist and Hellebrekers, 2003); ketamine does not produce visceral analgesia (NIH, 1991). For that reason, they are seldom used as the sole anesthetic in major surgery but instead are combined with other agents in a balanced anesthesia regimen (NIH, 1991). In these cases, drugs with different pharmacological effects are used in combination to produce a surgical anesthesia; for example, fentanyl and midazolam may be used in concert or ketamine and medetomidine may be combined (Hedenqvist and Hellebrekers, 2003). Using local anesthetics in combination with a general anesthetic is another example of a balanced anesthesia regimen, the benefit being that local anesthetics can reduce the need for general anesthesia and side effects associated with higher doses of general anesthesia (Gordon et al., 2002).

Analgesia is the inability to feel pain; an analgesic drug relieves pain but does not cause a loss of awareness. Analgesics include opioid drugs, such as morphine, and NSAIDs, such as aspirin and ibuprofen (NRC, 1992). There is evidence that surgery (or tissue injury) induces sensitization of central neural function, causing nociceptive inputs from the surgical wound to be perceived as more painful (hyperalgesia) than they would otherwise have been, and causes innocuous inputs to give rise to pain (allodynia). Studies have shown that preemptive analgesia (such as opiates, local anesthetics, or NSAIDs) prevents this sensitizing, reducing postoperative pain intensity and decreasing postoperative analgesic requirements for periods much longer than the duration of action of the preemptively administered analgesic (Coderre et al., 1993). Researchers should be encouraged to preemptively use analgesics.

Sedatives and anxiolytics may be used for the relief of non-pain-induced distress. They are often combined with analgesics to produce a state free of pain and distress—for example, in the management of postsurgical pain or pain associated with disease—and are also useful for restraint during minor procedures (NIH, 1991; NRC, 1992).

Systemic paralysis is commonly used in neuroscience experiments. These experiments require that the animal be paralyzed with a neuromuscular blocking agent to prevent movement, such as movement of the ocular muscles during visual experiments. Neuromuscular blocking agents are used only in fully anesthetized animals (NRC, 1996). They do not interact substantially with anesthetics and analgesics, but they leave an animal unable to respond behaviorally to pain or distress. That can make it difficult to evaluate the depth of anesthesia and the adequacy of analgesia, so other signs of pain or distress must be used, such as lacrimation, salivation, reactivity of heart rate and arterial blood pressure to noxious stimuli, or electroencephalographic recordings (NIH, 1991). Such signs are not adequate singly, but in combination they can provide valuable information about an animal's physiologic status (NIH, 1991). In addition, care should be taken to ensure that the animal has recovered control of respiration and locomotion before it is returned to the home cage. A detailed discussion of monitoring paralyzed animals can be found in Chapter 5.

It is important to confer with a laboratory-animal veterinarian to develop an adequate anesthetic and analgesic regimen. In fact, the AWRs states that a veterinarian be consulted during the planning of any procedure that could cause pain in animals (AWR 2.31(d)(1)(iv)(B)). Many resources are available to help the investigator and laboratory-animal veterinarian to develop a balanced anesthetic and analgesic regimen (Deyo, 1991; Flecknell, 1996, 1997; Hillyer and Quesenberry, 1997; Kohn et al., 1997; NRC, 1992; Rosenberg, 1991; Smith and Swindle, 1994; Stoelting, 1999).

## Asepsis and Physical Environment

Among the more problematic *Guide* recommendations for reviewers of proposed neuroscience protocols are those pertaining to physical environment and asepsis during surgery. The *Guide* states:

> In general, unless an exception is specifically justified as an essential component of the research protocol and approved by the IACUC, nonrodent aseptic surgery should be conducted only in facilities intended for that purpose [p. 62].
>
> For most rodent surgery, a facility may be small and simple, such as a dedicated space in a laboratory appropriately managed to minimize contamination from other activities in the room during surgery [p. 78].

The AWRs state:

> All survival surgery will be performed using aseptic procedures, including surgical gloves, masks, sterile instruments, and aseptic techniques. Major operative procedures on non-rodents will be conducted only in facilities intended for that purpose which shall be operated and maintained under aseptic conditions. Non-major operative procedures and all surgery on rodents do not require a dedicated facility, but must be performed using aseptic procedures. Operative procedures conducted at field sites need not be performed in dedicated facilities, but must be performed using aseptic procedures [AWR 2.31(d)(1)(ix)].

The *Guide* further states:

> The species of animal influences the components and intensity of the surgical program. The relative susceptibility of rodents to surgical infection has been debated; available data suggest that subclinical infections can cause adverse physiologic and behavioral responses (Beamer, 1972; Bradfield et al., 1992; Cunliffe-Beamer, 1993; Waynforth, 1980, 1987) that can affect both surgical success and research results [p. 63].

Many neuroscience procedures can be performed in full compliance with the AWRs and *Guide* recommendations for asepsis and the physical environment. However, if a survival surgical procedure requires the use of specialized equipment, facilities, or substances, performing it in a manner that complies fully with all recommendations can be impractical or impossible (NIH, 1991). Even when full compliance is not possible, most aspects of the recommendations can be met, such as the use of sterile surgical gloves, gowns, caps, and face masks; the use of sterile instruments; aseptic preparation of the surgical field; and appropriate postsurgical care.

IACUCs may receive requests to conduct survival neuroscience procedures in a modified laboratory setting to meet the scientific needs of an experiment, for example, when the experiment requires specialized equipment that cannot be sterilized or moved into a dedicated surgical facility. Both the AWRs and the *Guide* allow for such exceptions when "an acceptable justification for a departure is presented in writing" (AWR 2.31 (d)(1)) or "an exception is specifically justified as an essential component of the research protocol" (*Guide*, p. 62). Before granting such an exception to the AWRs and *Guide* recommendations, an IACUC should consider the extent to which animals will be susceptible to increased risk of infection. In particular, the IACUC should carefully review the various safeguards that will be used to minimize the risk. Examples of the safeguards are aseptic preparation of a separate area of the laboratory in which the surgery will be conducted; the use of aseptic surgical attire, instruments, and supplies; and aseptic preparation and maintenance of the surgical field during the procedure (NIH, 1991). Maintaining equipment that cannot be moved or sterilized under plastic or equivalent cover when not in use is encouraged to decrease any potential contamination of the equipment.

Several things can be done to make a laboratory setting more suitable for major survival surgery. The room should be free of unnecessary equipment. In some situations, a large, general-purpose laboratory can be partitioned to isolate a smaller surgical area. The room in which surgery is to be performed must be sanitized immediately before each procedure. The walls, ceiling, and floor should have smooth surfaces that are impervious to moisture and easily cleaned (NIH, 1991).

A decision to allow major survival surgery to be performed in a modified laboratory setting should be contingent on the development of a set of stringent postsurgical monitoring and reporting procedures. For example, the IACUC may approve an exception to the *Guide's* recommendations subject to receiving a status report from veterinary staff on the health and welfare of animals during the postsurgical survival period, to ensure the efficacy of the various procedures proposed to mitigate the risk of postsurgical infection.

## Postsurgical Recovery Period

As noted in the *Guide*:

> The investigator and veterinarian share responsibility for ensuring that postsurgical care is appropriate. An important component of postsurgical care is observation of the animal and intervention as required during recovery from anesthesia and surgery. The intensity of monitoring necessary will vary with the species and the procedure and might be greater during the immediate anesthetic-recovery period than later in postoperative recovery [p. 63].

Temperature and hydration should be monitored, maintained, and recorded until recovery from anesthesia. Monitoring heart rate and respiratory rate may also prove useful. Animals should be monitored until it is determined not only that the animal is normothermic, but also that it can maintain a normal body temperature in the absence of supplemental heat (Waynforth et al., 2003). Small mammals, such as rodents, are particularly susceptible to hypothermia and must be monitored closely (Hedenqvist and Hellebrekers, 2003). Because the period of recovery from anesthesia is often a time of substantial physiologic change, policies and procedures that ensure adequate observation of the animal to facilitate prompt correction of problems should be implemented. According to the *Guide*:

> During the anesthetic-recovery period, the animal should be in a clean, dry area where trained personnel can observe it often. Particular attention should be given to thermoregulation, cardiovascular and respiratory function, and postoperative pain or discomfort during recovery from anesthesia. Additional care might be warranted, including administration of parenteral fluids for maintenance of water and electrolyte balance (FBR, 1987), analgesics, and other drugs; care for surgical incisions; and maintenance of appropriate medical records [pp. 63–64].
>
> After anesthetic recovery, monitoring is often less intense but should include attention to basic biologic functions of intake and elimination and behavioral signs of postoperative pain, monitoring for postsurgical infections, monitoring of the surgical incision, bandaging as appropriate, and timely removal of skin sutures, clips, or staples (UFAW, 1989, p. 64).

## PHYSICAL RESTRAINT

### Determining the Appropriate Restraint Procedure

Restraint has been characterized as a physiologic and psychologic stressor (Norman et al., 1994; Norman and Smith, 1992). The method of animal restraint used to achieve a particular objective in an experimental protocol should, according to the US Government Principles (IRAC, 1985), be chosen so as to minimize distress to the animal. And the *Guide* (p. 11) states that prolonged restraint should be avoided unless it is scientifically justified and approved by the IACUC. The *Guide* recommends that less restrictive methods be chosen when possible. The objectives of restraint should be clearly defined in the animal-use protocol. Specifically, the degree of restraint needed (head only, arms and head, whole body, and so on) will guide the method of restraint and the type of equipment used.

Personnel safety may also necessitate restraint of an animal. For example, neuroscience studies that require physical proximity of macaques and the experimenter should involve sufficient restraint of the primates to minimize the risk of handler exposure to B virus by scratch, bite, or splash. Similarly, a rabbit-restraint device has been described that secures the animal and immobilizes its feet to prevent scratching of the experimenter (Abell et al., 1995). Animals that have been habituated to a suitable restraint method will probably be less stressed and agitated during a procedure, thereby reducing risk of injury to the handler (Sauceda and Schmidt, 2000).

Newer restraint devices and techniques reduce distress and enhance the health and comfort of animals by taking into account their behavior and typical postural adjustments. For example, Binder (1996) describes a short-term mouse-restraint device that is based on the mouse's preference to seek shelter; the mouse reliably and voluntarily crawls forward so that its head is in an opening of a box while the experimenter gently restrains it by holding on to its tail. Such a restraint procedure is presumably less stressful for the animal, because the animal can express a normal coping behavior.

Observation of an animal during restraint is critical and should be thoroughly described in the animal-use protocol. Observation may be direct, such as through a viewing window (e.g., Ator, 1991), or indirect, such as by remote video. The frequency of animal observation may depend on the specific experimental procedure, but it may also be determined by the species of animal, the degree and duration of restraint, the type of restraint device or technique, the stage of training of the animal to the restraint, and the animal's degree of habituation to the restraint. Observation may include a physical examination, an evaluation of the animal's behavior, and/or an assessment of various physiologic measures, such as concentrations of cortisol/corticosterone, leuteinizing hormone (LH), testosterone, and blood glucose) (Flecknell and Silverman, 2000; Norman and Smith, 1992; Rogers et al., 2002; Wade and Ortiz, 1997). Monitoring should be frequent because restraint-device failures and unanticipated actions by the animal can sometimes place the animal in jeopardy. Detailed records should indicate the date, the time, the observer's name, and the observations made.

Occasionally, an animal will not adapt well to restraint. Criteria for the temporary or permanent removal of an animal from a study that requires restraint must be developed in advance of the study and be reviewed and approved by the IACUC. The development of physical or behavioral abnormalities should constitute a basis of a decision to temporarily or permanently remove an animal from a study.

## Methods of Physical Restraint

The two principal methods of physical restraint are manual and device-facilitated. In general, manual restraint is used for short-term procedures. Personal protective equipment (such as gloves) is often used to enhance worker safety by preventing bites, scratches, or contact allergy that can occur with some species (Egglestone and Wood, 1992).

Device-facilitated restraints can be used safely in some situations. Depending on the goals of the study, the equipment can facilitate short-term restraint (e.g., Abell et al., 1995) or be incorporated into the animal housing (e.g., Coelho and Carey, 1990). Innovative restraint equipment, such as slings and restraining boxes, has been used successfully without increasing stress (Flecknell and

Silverman, 2000). Sauceda and Schmidt (2000) have reviewed common restraint devices used with macaques.

### Conditioning an Animal to the Restraint Technique

An experimental procedure that requires performance or measurements of restrained animals can be successful only if the animal is not unduly stressed and is sufficiently habituated to the restraint so that it will attend to the task rather than focusing on the restraint itself. Thus, an initial investment of time to train the animal to accept the restraint is highly recommended, particularly for chronic or repeated restraint. In accordance with the AWRs (AWR 2.32), personnel working with restrained animals should be trained in using the equipment properly and in handling the animals safely while causing them minimal distress. Well-qualified personnel will have a sound understanding of when restraint should be suspended or stopped if it compromises animal welfare.

The period required for habituation of an animal to a restraint technique varies. For a single brief period of restraint, habituation may not be critical for obtaining valid data. In some studies, providing social animals with companionship during the restraint period may reduce restraint-related stress. For example, Fleischman and Chez (1974) chair-restrained baboons as pairs to reduce anxiety. Restraint can result in various physiologic changes in animals (e.g., Bush et al., 1977; Gartner et al., 1980), so a substantial period of habituation may be required to obtain valid data. The period of training depends on the species and the animal's experience and behavior; animals will habituate to the procedure better if the equipment and handling procedure are species-appropriate, sized correctly for the animals, and otherwise adjusted to maximize the animals' comfort. The habituation period is especially critical for studies that require more restrictive restraint (NIH, 2002). Maintenance of physiologic and behavioral measures within normal limits during restraint suggests that an animal is well adapted to the restraint, as do voluntary movement of the animal into the restraint equipment and performance of the requisite task (NIH, 2002). For example, Wade and Ortiz (1997) demonstrated that well-habituated monkeys had no rise in urinary cortisol associated with restraint.

### Potential Consequences of Restraint

The correct use of restraint can facilitate the collection of accurate research data. However, the inappropriate application of restraint can adversely affect health, physiologic measurements, and behavior. An animal that has had an adverse experience during restraint may be more difficult to use in the future because of increased anxiety resulting from memory of the experience. The development of ulceration on the ischial callosities of some primates as a result of chronic restraint has been reported (e.g., Wade and Ortiz, 1997). Restraint has also been shown to inhibit LH and testosterone secretion in male macaques and LH secretion during the follicular phase of the menstrual cycle of macaques, resulting in reduced fertility (Norman et al., 1994; Norman and Smith, 1992). Restraint that causes stress activates the hypothalamic-pituitary-adrenal axis. Thus, the potential scope of adverse effects on an animal bears careful consideration. Under all circumstances, the minimal restraint feasible should be used—such as a tether in lieu of chair restraint, or a shorter period of restraint—and the availability of adequate alternatives to the restraint should be assessed (Flecknell and Silverman, 2000).

### Prolonged Physical Restraint

The *Guide* states, "prolonged restraint, including chairing of nonhuman primates, should be avoided unless it is essential for achieving research objectives and is approved by the IACUC" (p. 11). Although prolonged restraint has not been defined, as the duration of restraint increases, a concomitant increase in attention should be given to alternatives to restraint, the health and well-being of the animal, and endpoint criteria for the restraint. The AWRs direct the IACUC to review procedures to avoid or minimize animal discomfort, distress, and pain and direct the investigator to consider alternatives to procedures that may invoke more than momentary or slight pain and/or distress (AWR 2.31 (d)(1)(i,ii,iii,iv) and APHIS/AC Policy 11). The AWRs go on to state that in instances where long-term (greater than 12 hours) restraint is required, a nonhuman primate must be provided the opportunity daily for unrestrained activity for at least one continuous hour during the period of restraint, unless continuous restraint is justified for scientific reasons and approved by the IACUC (AWR 3.81 (d)).

## FOOD AND FLUID REGULATION

Neuroscience-related protocols occasionally require the regulation of animals' food or fluid intake to achieve a specific experimental goal. The regulation process may entail *scheduling* of access to food or fluid sources so an animal consumes as much as desired at regular intervals, or *restriction*, in which the total volume of food or fluid consumed is strictly monitored and controlled. As stated in the *Guide,* "the least restriction that will achieve the scientific objective should be used" (p. 12). Research protocols that use food or fluid regulation can be divided into at least three main categories: studies of homeostatic regulation of energy metabolism or fluid balance, studies of the motivated behaviors and physiologic mediators of hunger or thirst, and studies that regulate food or fluid consumption to motivate animals to perform novel or learned tasks (Toth and Gardiner, 2000).

In studies of homeostatic regulation, the manipulation of food or fluid availability would be predicted to directly influence a dependent variable that is being measured as a specific aim of the experiment, for example, food restriction leads to neurally mediated hormone release.

In contrast, regulation of food or fluid is commonly used as motivation in experiments that require animals to perform a behavioral task with a high degree of repeatability (Toth and Gardiner, 2000), but the food or fluid consumption is not the experimental variable. In those studies, food and fluid regulation is used to motivate the animals to perform a specific behavioral task for a food or fluid reward; regulation of food or fluid outside the experimental session ensures response reliability to the food and fluid reward in each session (NIH, 2002). That allows the investigator to elicit and monitor the same movement repeatedly, to present the sensory stimuli under highly controlled conditions, and to obtain physiologic discriminations from the animal. For example, water-regulated monkeys may be trained to press a button for a juice reward, while the investigator measures the effect on neuronal firing rates. In conditioned-response experiments, (for example, a monkey may be conditioned to associate a light with a fluid reward), consideration should be given to whether the use of highly preferred food or fluid as positive reinforcement can be used instead of restriction.

Fluid reward is preferable to food reward in some types of experiments. For example, studies that monitor neuronal activity in the brain may require the minimization of jaw or head movement to avoid displacing a microelectrode from its position. Because fluid rewards can be delivered through a tube positioned near the animal's mouth and tongue, they offer a particular advantage: licking and swallowing a fluid reward are much less disruptive to the neuronal recordings than chewing or crunching movements of the teeth or jaws that accompany the consumption of food rewards (NIH, 2002).

Fluids offer additional experimental advantages. They can be easily delivered in small quantities, maximizing the number of trials that can be executed before satiation of the animal. In contrast with food rewards that require chewing before swallowing, fluids are quickly consumed, reducing the intertrial interval—an important advantage when an animal must perform a behavior hundreds or even thousands of times in an experimental session to allow for statistical analysis.

In other studies, there may be disadvantages to using fluid rewards. For example, milk and juice require more extensive cleaning than water or solid food if spilled on the experimental apparatus. Milk and juice are also more susceptible to rapid spoilage and require frequent assessment or replacement.

In designing and evaluating an animal-use protocol that proposes to regulate access to food or fluid to facilitate operant training, the following questions should be considered:

- What type food or fluid regulation is most appropriate for meeting the specific goals of the experiment?

- Do alternative procedures exist that would facilitate the generation of the desired behavior without food or fluid regulation, or is food or fluid regulation the best option?
- What is the proposed schedule of food and fluid access, and does it allow periodic ad lib access to food and fluid?
- What is the proposed schedule for monitoring, so adverse effects will be recognized quickly.
- Is laboratory chow or fluid the only item to be offered, or will other foods or fluids be considered?
- What are the endpoints for intervention with supplemental feeding or hydration?

The development of animal protocols that involve the use of food or fluid regulation requires the determination of three fundamental details: the necessary level of regulation, the potentially adverse consequences of regulation, and methods for assessing the health and well-being of the animals. Consideration of each of those details facilitates the establishment of interventional endpoints to maintain the animals' health and well-being.

### Food Regulation

The *Guide* states that when the experimental situations require food or fluid regulation, at least minimal quantities of food and fluid should be available to provide for development of young animals and to maintain long-term well-being of all animals. Regulation for research purposes should be scientifically justified and approved by the IACUC. A program should be established to monitor physiologic or behavioral indexes, including criteria (such as weight loss or state of hydration) for temporary or permanent removal of an animal from the experimental protocol (NRC, 1996, p. 12). Reducing an animal's body weight by 15–20% (compared with cage-matched controls) is commonly the goal of food regulation (NIH, 2002).

In general, the total caloric intake of a food-regulated animal is 50–70% of that associated with ad libitum feeding (Bucci, 1992). In some cases, however, the attending veterinarian may determine that an animal needs to be removed from a study for health or behavioral reasons even if it has not reached that weight loss. In addition, species, strain, and individual differences may influence what is considered an acceptable amount of weight loss.

Special attention should be given to ensuring that the diet meets the animal's nutritional needs (New York Academy of Sciences and Ad Hoc Committee on Animal Research, 1988) unless the scientific needs of the research protocol necessitate otherwise. Caloric restriction must not produce unintended nutritional imbalances. In some cases, it might be necessary to increase the concentration of selected nutrients to provide the same nutrition as provided to animals fed ad libitum (NRC, 1995). Typically, the restricted diet contains proportionate decreases in energy sources (protein, fat, and carbohydrates) rather than a reduction in only one source of energy (Bucci, 1992).

The AWRs state that procedures involving more than momentary or slight aversive stimulation that is not relieved with medication or other acceptable methods should be undertaken only when the objectives of the research cannot be achieved otherwise. APHIS/AC's Policy 11, "Painful Procedures," lists "food or water deprivation beyond that necessary for normal presurgical preparation" as a procedure that may cause pain or distress. The IACUC should closely evaluate the pain-distress categorization of animals that are food-restricted in accordance with APHIS/AC Policy 11.

In some rats, a decrease of 20% in baseline body weight within 1 week is associated with increased serum corticosterone, which may reflect the physiologic response to caloric restriction, and with substantially greater freezing behavior in the open field test, which may be a stress response (Heiderstadt et al., 2000). Generally, it is recommended that animals be gradually reduced to a target weight and acclimated to the feeding schedule over some period, such as several weeks, to mitigate the stress response. The stress response associated with a rapid reduction in body weight can also be relieved by following the reduction with a diet designed to maintain the rat at 80% normal body weight, compared with a rat fed ad libitum. (Heiderstadt et al., 2000).

### Mechanisms for Mediation of Food Consumption on an Ad Libitum Diet

There is individual variability in food intake and adult body weight (Toth and Gardiner, 2000). Food intake may be regulated more by satiety than by hunger (Stricker, 1984). Physiologic signs of satiety include gastric distention and increases in insulin secretion and metabolic processes in the liver (Toth and Gardiner, 2000). Food consumption is influenced by palatability and accessibility (Collier et al., 1972; Peck, 1978; Rolls et al., 1983). The amount of work an animal must do for food also influences the amount consumed; increases in workload reduce consumption to about 50% of control consumption (Collier, 1989; Collier et al., 1972; Nicolaidis and Rowland, 1977). Rowland et al. (1996) have summarized the various endogenous and exogenous regulators of food intake.

### Determination of Minimum Caloric Consumption

A sound approach to developing a food-regulation protocol requires information about the minimum caloric requirements of animals. A general method of assessing caloric consumption is to require an animal to work for all its food under different reinforcement schedules (Collier, 1989; Nicolaidis and Rowland, 1977). Caloric needs and consumption vary with the life stage of the animal, such as growth, maintenance, gestation, and lactation; activity level; environmental conditions, including social housing and thermoregulatory demands; and circadian rhythm. However, some animals consume more than is necessary to meet their metabolic needs (Toth and Gardiner, 2000). In some species, preventing coprophagy (ingestion of feces) results in an increase in nutritional requirements. Investigators should consider those factors in determining nutritional requirements related to different circumstances.

### Food-Regulation Design

Two common methods are used for regulating the food intake of animals to motivate them to perform tasks. The first method restricts the amount of time available to the animals to eat, and the second restricts the amount of food available. When the former method, referred to as meal feeding, is used, the growth curve of animals generally remains below that of animals fed ad libitum; meal feeding also results in a different pattern of drinking, in which most of the volume is consumed during the meal period. When the latter method, referred to as restricted feeding, is used, a lower body weight also occurs. Both types of regulation cause a temporary increase in food efficiency in young rats, and the reduced intake is correlated with a reduction in the animal's relative fat content and an increase in relative water content (Brownlow et al., 1993). However, rats on a restricted diet that was provided as two meals, rather than one, had substantially more adipose tissue. Food-restricted rats normalize their rate of weight gain (relative to control animals), although they will not achieve the mean body weight of satiated animals. The choice of regulation method should be based on the goals of the study and the behavior of the animal.

### Species- and Strain-Specific Considerations

As there are species- and strain-related differences in feeding behaviors, there can also be variation in responses to food scheduling, so the amount and pattern of food regulation necessary to induce animals to perform tasks varies. Rats and mice are meal-eaters rather than nibblers (Classen, 1994a), and rats have a circadian rhythm of feeding (Classen, 1994b) and are more likely to eat during the dark cycle (Lima et al., 1981); these factors may affect an animal's response to a particular food schedule (Bellinger and Mendel, 1975). Rats adjust to reasonable food-restriction regimens; however, several studies indicate that mice are less resilient and that food restriction can compromise their well-being (Nelson et al., 1973). Intolerance of food restriction is aggravated by single housing and by feeding during the light cycle (Hotz et al., 1987; Van Leeuwen et al., 1997). Thus, the species and the sensory environment can affect the physiologic response to food scheduling.

Depending on the species, food regulation can have secondary effects on research. For example, in rats, scheduled access to food can result in an increase in the self-administration of drugs that are perceived by the subject as reinforcing agents (Laties, 1987). Diet-restricted rats may also be more agitated during restraint (Albee et al., 1987). Rats can exhibit hemoconcentration that is directly related to the degree of food restriction (Levin et al., 1993). Other physiologic alterations observed in meal-fed rats include reduced white-cell counts, reduced platelet counts, reduced serum protein concentration, increased serum bilirubin, decreased cholesterol (females only), imbalances in electrolytes, reduced hematopoietic tissue in sternal bone marrow, and, in severely restricted rats, bone marrow necrosis, thymic atrophy, and mild testicular degeneration (Levin et al., 1993). Nonhuman primates that are fed a calorie-restricted diet have a reduced bone mass, slightly lower body temperature, and increased glucose tolerance and insulin sensitivity (Roth et al., 2000).

### Influence of Circadian Rhythm

The circadian rhythms of several physiologic and behavioral variables are affected when access to food is limited to particular times of the day. Rats that are meal-fed shift their activity pattern relative to the timing of meal presentation in such a way that periods of quiescence and activity are anchored to the time of day of feeding (Classen, 1994b). Factors that are affected in the mouse include the circadian rhythm of plasma corticosterone concentrations and core body temperature (Classen, 1994b).

## Fluid Regulation

### Physiologic Mechanisms of Fluid Consumption

Three main physiologic stimuli mediate thirst, fluid consumption, and hydration in normal animals maintained on an ad libitum fluid-consumption schedule (Rolls and Rolls, 1981; Rolls et al., 1980; Stricker and Verbalis, 1988; Toth and Gardiner, 2000; Wood et al., 1982). The first is cellular dehydration, which may result from inadequate water consumption, excessive renal or evaporative water loss, or ingestion of excessive quantities of solutes, such as sodium. Fluid consumption maintains osmotic balance by reducing the extracellular solute concentration and thus allowing fluid to move back into the cells to restore intracellular fluid volume. The second is hypovolemic thirst, which occurs when fluid is lost from the blood, as occurs during dehydration. The hypovolemic status of an animal can be ascertained by measuring hematocrit or plasma protein, which increase when fluid is lost from the plasma (Toth and Gardiner, 2000). The restoration of normal plasma volumes requires solute ingestion concurrent with fluid consumption because fluid consumed in the absence of solutes moves into the intracellular compartment. The third stimulus is the hormone angiotensin, which stimulates drinking in some physiologic states.

Mechanisms for fluid conservation are activated in the kidneys during cellular dehydration or hypovolemia. The progression of dehydration over consecutive 12-hour periods is nonlinear because mechanisms for fluid conservation are invoked progressively to retard water loss (Toth and Gardiner, 2000). There are species-specific differences in the degree of fluid loss from cellular and plasma compartments (Rolls et al., 1980), which affect the degree of cellular dehydration or hypovolemia that occurs because of fluid deprivation. Especially with small animals (NIH, 2002), there is a potential for dehydration as a result of fluid loss from both intracellular and plasma compartments (Toth and Gardiner, 2000). In rats, dogs, and monkeys, cellular dehydration and hypovolemia are the primary physiologic variables that mediate fluid consumption (Fitzsimons, 1998), and species-specific differences in drinking after fluid deprivation are apparent.

### Determination of Minimum Fluid Consumption

It is difficult to specify minimum fluid requirements for the various animal species, because there is a dearth of evidence in the scientific literature. That contrasts with the growing literature on the health consequences of caloric restriction.

Physiologic needs for water are influenced by many factors, including the water and electrolyte content of the diet, the ambient temperature and humidity, and exercise. Fluid consumption is also influenced by nonhydration variables, such as habit, social factors, palatability, and ease of access to fluids. Those variables tend to increase the average daily consumption of fluids to more than is necessary to maintain homeostasis (Nicolaidis and Rowland, 1975; Rowland and Flamm, 1977). Attempts to estimate a socially housed animal's daily fluid needs on the basis of "everyday experience" are likely to lead to inflated estimates because much fluid consumption is motivated by social or other variables rather than by hydration needs (Mountcastle, 1980).

Fluid maintenance requirements vary markedly across species; fluid maintenance requirements range from 35 to 140 mL/kg of body weight (BW) per day (Aiello, 1998a; Kirk and Bistner, 1985; NRC, 1995; Wells et al., 1993; Wood et al., 1982). There can also be a wide range of ad libitum consumption levels within a species; for example, daily fluid consumption in nonhuman primates has been reported at 75 mL/kg BW (Kerr, 1972; Wayner, 1964), 90 mL/kg BW (Wayner, 1964), and 110 mL/kg BW (Evans, 1990). There can also be a wide range within a strain; for example, daily ad libitum fluid consumption in 3 different Sprague-Dawley rat colonies was reported at 80 mL/kg BW, 105 mL/kg BW, and 125 mL/kg BW (Wells et al., 1993). There can even be significant gender differences; for example, daily ad libitum fluid consumption in male golden hamsters was reported at 50 mL/kg BW, while female golden hamsters consumed 140 mL/kg BW (Fitts and St Dennis, 1981). Consequently, assessing the ad libitum fluid consumption for each fluid-regulated experimental animal might be an important step in ensuring the health and well-being of the animal.

However, voluntary fluid-consumption levels in a laboratory setting might not be equivalent to the animal's minimal fluid-requirement levels. Limited availability of fluid is a common determinant of consumption in natural settings, and physiologic and behavioral mechanisms have evolved to enable animals to adapt to the limitation. For example, rats and monkeys quickly learn to consume much, if not all, of their daily fluid needs in a short, restricted period (reviewed by Evans, 1990). Species that drink from watering sites only once per day invoke homeostatic mechanisms to control urine output in relation to their hydration state (Toth and Gardiner, 2000). Mammals may also use torpor to adapt to the dry season in their natural habitat (Schmid and Speakman, 2000). Thus, it is difficult to designate specific minimum fluid needs, because requirements may vary with species, strain, environment, efficiency of fluid-saving mechanisms, and so on. IACUCs, veterinarians, and researchers should take into account the possibility that laboratory animals can be adequately physiologically sustained with less fluid than they would voluntarily consume.

At the start of a new research protocol involving restricted or altered access to fluid, the amount of fluid consumed, body weight, and a hydration assessment should be recorded daily for each animal, as individual animals may manifest physiologic and behavior differences. Those data will help in refining the protocol and evaluating the adequacy of access to fluid. In evaluating the adequacy of access to fluid, each animal should be evaluated individually to determine how it is adapting to the imposition of restricted or altered access. For example, if an animal attains and then maintains a new body weight, it could suggest successful adaptation even if the new weight is below the weight recorded during ad libitum access to fluid.

### Fluid-Regulation Design

When fluid regulation is selected as a behavioral motivator, access to fluid outside the experimental setting has to be regulated to motivate performance of the rewarded behavior (NIH, 2002). Generally, fluid regulation is patterned after one of two designs. In "fluid restriction," animals are given access to a metered volume of fluid per day and may consume that volume over any length of time. In "fluid scheduling," the experimenter determines the time of day during which the animal has access to fluid, but the duration of drinking and the volume consumed are determined by the behavior of the animal. For example, in many behavioral protocols, animals are given continual access to fluid for as long as they continue to perform a task. Because food and fluid generally are not freely available in the wild and some effort (foraging) is required to obtain them (NIH, 2002), such scheduling designs may model the effort expenditure necessary to obtain food and fluid in the wild.

For both types of fluid regulation, animals generally should be given free access to fluid for some period on days when experimental sessions are not scheduled, unless scientifically justifiable reasons preclude such fluid supplementation (NIH, 2002).

When developing a restriction design to motivate an animal to perform a task, the main consideration is determining what level of restriction is necessary to achieve the desired performance. Generally, the more complicated the task, the more stringent the restriction protocol needs to be. For example, in a study of water-restricted rats, where the rats were required to bar press to obtain their daily allotment of water (Collier and Levitsky, 1967), mild restriction (rats receive 75% of their average ad libitum intake) resulted in poor performance while a more stringent restriction (rats receive 32% of their average ad libitum intake) resulted in maximal performance. This is why fluid restriction levels used in one study may not provide adequate motivation for learning or performing other more demanding tasks (Toth and Gardiner, 2000). However, the most severe restriction is not always necessary for achieving maximal performance. In this same study (Collier and Levitsky, 1967), bar-press rates were similar when water was restricted to 32%, 42%, or 56% of average ad libitum intake.

The health implications of fluid regulation have been of concern, although even chronic restriction schedules have not been found to cause physiologic impairment of animals that are adapted to the restriction and receive enough fluid to replenish daily losses (Toth and Gardiner, 2000). For example, the consequences of using fluid scheduling to motivate lever-pressing behavior have been examined in rats deprived of fluid for 7, 14, or 21 hours/day for 3 months. The animals showed no observable adverse effects compared with ad libitum controls with respect to weight loss, organ and tissue appearance at necropsy, hematologic examination, or clinical chemical analysis (Hughes et al., 1994).

## Species- and Strain-Specific Considerations

The development of fluid-regulation schedules should include some consideration of species variations in fluid consumption behaviors. Some species consume fluid intermittently, sometimes only once per day, but others consume smaller quantities more frequently (Rolls et al., 1980). Efforts should be made to match an animal's typical watering schedules with circadian variables, because the risk to animals on fluid regulation is reduced if periods of access and total amounts available are appropriate to the species (NRC, 1995; Toth and Gardiner, 2000). In addition to the behavioral aspects of fluid consumption, relationships between fluid intake and food intake should be considered. Food ideally is provided at close to the same time as daily fluid provision (for example, after the experimental session):

> The concurrent availability of water and food incurs two benefits. First, fluid intake promotes food intake, thereby reducing the likelihood of dehydration-

related anorexia. Second, the consumption of food associated with water allows animals to consume solutes that will help retain water in the circulation, correct volume deficits, and avoid excessive hemodilution that will cause urinary excretion of the ingested water (Toth and Gardiner, 2000).

## Other Influences on Fluid Homeostasis

In some situations, fluid reinforcers (such as fruit juice) are used because they may maintain behavioral performance when access to fluid is restricted; for example, some monkeys prefer fruit juice when performing long behavioral sessions in which many reinforcements are delivered (NIH, 2002). Investigators, veterinary personnel, and IACUCs should consider and monitor for any potential physiologic ramifications of total substitution of solute-containing fluids for water in a fluid-restricted protocol. Sweetened milk or juices may be unfavorable choices for use in a long-term study in which an animal will participate for many months or years, because of the potential for dental caries (NIH, 2002).

Provision of treats, such as fruits or vegetables, is recommended when appropriate to provide variety and nutritional balance to an experimental animal's diet (NRC, 1996). The water content of these dietary supplements can be difficult to estimate, so their potential contribution to hydration should not be considered in determining the minimal ration of fluids to be given to the animal (see Pennington et al., 1998, for data on water content of fruit and vegetable supplements). However, investigators, veterinary personnel, and caretakers should be aware of the potential need for restriction or substitution of supplemental food items in fluid-regulated animals.

## Variability Between Individuals

When presented with the homeostatic challenge of dehydration, animals can respond by conserving water and excreting concentrated urine (physiologic regulators) and/or by drinking more fluid and excreting dilute urine (behavioral regulators) (Kanter, 1953; Toth and Gardiner, 2000). Animals on fluid restriction or scheduling protocols may implement different compensatory mechanisms to different extents. Animals that are physiologic regulators may be problematic when used in behavioral studies, because they often tend to accommodate to the consumption of a minimal volume of fluid by excreting more concentrated urine instead of consuming more fluid (Toth and Gardiner, 2000). In contrast, the behavioral regulator tends to modify its behavior during the experimental task to obtain more fluid as a reward. In both instances, nonhuman primates often supplement fluid consumption by licking water from cages after washing. Therefore, the assessment of each animal on a fluid-regulation protocol is prudent.

## Training Considerations

The difficulty of the behavioral task that an animal must learn and the goals of the experiment often influence the degree to which the animal must be motivated to perform the task and thus the degree of fluid regulation that is necessary. When possible, palatable rewards rather than regulation should be used to motivate behavior. However, if fluid regulation is determined to be the preferred method of motivating a particular behavior, consultation with veterinary personnel and a review of recent literature regarding animal training may be appropriate (NIH, 2002).

In training of a naïve subject to perform a new task with a fluid regulation, gradual introduction to the concept that fluid availability is restricted or context-dependent (for example, earned while in the experimental apparatus) is important (Toth and Gardiner, 2000). After the animal has experienced the absence of ad libitum fluid, its motivation to learn or perform tasks to earn fluid usually increases. The degree of restriction may require periodic adjustment to generate adequate motivation to learn or perform difficult phases of a task (Toth and Gardiner, 2000). However, the restriction often may be reduced after the animal learns the task and becomes proficient at it. As noted in *Methods and Welfare Considerations in Behavioral Research with Animals* (NIH, 2002):

> When the study begins, be prepared to consider and address a range of behavioral, environmental, or equipment-related variables that might hinder training or disrupt performance. Inexperienced personnel may presume that a source of problems in training or maintaining a food- or fluid-motivated behavior is that the restriction is not strict enough (or, in some cases, that it is too strict). The other types of variables that should be considered first, however, are equipment malfunctions, programming errors, task criteria that are raised rapidly or set too high for the animal's level of training, illness, or nonprogrammed water restriction (in the case of food-motivated behavior).

Furthermore, experimental animals, like humans, may have deficits, such as myopia, that impair performance on tasks because of perceptual limitations. "In all circumstances, careful monitoring of animals under food or fluid control is necessary every day to avoid additional nonprogrammed restriction" (NIH, 2002).

### Assessment of Animal Subjects as Individuals

The previous paragraphs emphasize that animals on food or fluid regulation schedules are individuals whose performance is likely to vary from day to day. Variations between individual animals in performance on a given task are also expected. The differences between individuals and even within an individual during different phases of an experiment may occasionally necessitate some adjustment of food or fluid scheduling to maintain homeostatic balance and achieve the desired experimental goals. The frequency of observations should therefore be adjusted according to how fast an animal can be compromised in the experimental situation. Diligent record keeping on daily food or fluid volume consumed, hydration status, appearance, general affect, experimental performance, and routine weighing are reliable for identifying changes in behavior patterns. Those records should be reviewed regularly and kept readily accessible to the veterinary staff and others who may have a need to evaluate them, such as the IACUC during its semiannual inspections. The need for intervention or reassessment of the hydration needs of an experimental animal can thus be recognized before adverse physiologic consequences develop.

## Methods of Assessing Nutrition and Hydration

A system of daily monitoring procedures is essential for animals that are food or fluid regulated. Records should be kept of the amount of food or fluid earned in the behavior task as well as any supplements given. Careful observation of the animal's behavior and regular clinical monitoring of the animal's health are critical to ensuring successful application of food or fluid regulation (NIH, 2002).

Clinical monitoring should include assessments of the nutritional and hydration state of experimental animals whose access to food or fluid is regulated. There are various methods for assessing a food- or water-regulated animal and reliance on a single variable is discouraged. Instead, investigators, caretakers, and veterinary staff should use several methods concurrently to ensure the health and well-being of the animals. Variables that can be monitored to assess the nutritional or hydration status of experimental subjects include the following.

*Weight and food intake*

Experimental animals on food or fluid regulation should be weighed several times a week, ideally before experimental sessions (NIH, 2002). Some accommodations in the frequency of weighing may be necessary if experimental animals require sedation or anesthesia to be weighed (NIH, 2002). Often, animals can be trained to cooperate with the procedures. Despite conditioning, however, the process of weighing may be very stressful to some experimental animals. In such cases, an animal can be weighed less frequently, and other reliable methods of hydration monitoring can be used.

Aside from daily fluctuations in weight due to fluid gain and loss, animals on fluid regulation may lose weight as a result of decreased food ingestion. Using a percentage weight-loss criterion during fluid deprivation as an endpoint for determining when an animal should be removed from a fluid restriction paradigm and their fluid requirements reassessed can indicate not only a proper level of motivation, but also health (Bolles, 1975). The amount of food consumed by experimental animals is a good measure of general health and of hydration status and should be monitored by caretakers or investigators, or both. Persistent decreases in food consumption should be brought to the attention of appropriate veterinary personnel.

*Skin turgor*

The texture and elasticity of skin are important indicators of an animal's hydration status. Ordinarily, dehydration will cause a slow return of the skin to its normal position after it has been lifted. However, that characteristic is less reliable in obese animals because their skin tends to maintain its elasticity even in the presence of dehydration (Kirk and Bistner, 1985).

*Solid and fluid waste output and moistness of feces*

As part of the long-term adaptation to fluid restriction, healthy animals produce concentrated urine and feces that are less moist than normal (Toth and Gardiner, 2000). Regular observation of the quantity and qualities of the excrement produced by an animal on a fluid or food regulation provides information about both hydration status and physiologic compensation for fluid regulation.

*General appearance and demeanor and quality of fur and skin*

Investigators and veterinary personnel share the responsibility for observing behavior, general appearance, and demeanor of experimental animals, which can be valuable indicators of their health status. For example, dry mucus membranes and sunken eyes are indications of dehydration (Aiello, 1998a). If signs indicate that an animal is developing problems related to dehydration, hemoglobin content or hematocrit and blood urea nitrogen can be measured to determine its physiologic status.

A plan of action, complete with endpoints for therapeutic intervention, should be considered when the experimental animal protocol is being developed. The plan should include standard operating procedures to be used if an animal develops diarrhea or vomiting that would prompt the return to an ad libitum fluid or food schedule and application of a schedule for veterinary evaluation to prevent serious health consequences due to dehydration or malnutrition.

## GENETICALLY MODIFIED ANIMALS

Genetically modified animals have induced mutations that are human-made alterations in their genetic code. The generic phrase *genetically modified* includes both transgenic and targeted mutations that are created to study the expression, overexpression, or underexpression of a specific gene (ARENA-OLAW, 2002). A transgenic animal has genes from another organism or species incorporated into its genome, whereas an animal with a targeted mutation has had the coding sequence of a gene in its own genome altered. For a genetic modification to be useful in research animals, the introduced or altered gene must be transmitted to the offspring. Most induced mutations have been made in laboratory strains of mice (*Mus*) or rats (*Rattus*). Although mice are used as examples in the following discussion, the general considerations are applicable to induced mutants of any species (ARENA-OLAW, 2002).

Genetically modified animals are used to test hypotheses in several ways: the phenotype of the modified animal is evaluated to determine the pathogenesis of disease or gene influences on development, the modified animal is used to test interventions to treat its condition, or the animal is used as a tool to study the pathogenesis of other conditions.

## Transgenic Animals

A *transgenic* animal has exogenous (foreign) deoxyribonucleic acid (DNA) inserted into its cells. Typically, transgenic animals are created by the "pronucleus method," in which complimentary deoxyribonucleic acid (cDNA) made from specific messenger ribonucleic acid (mRNA) is inserted into cells by using microinjection, electroporation, or nonpathogenic viruses. Each of those methods has been used to insert new DNA into the pronucleus of a fertilized mouse egg to create viable transgenic mice. The manipulated fertilized eggs may be cultured in vitro for several days before they are surgically implanted into the oviducts or uterus of pseudopregnant female mice. The successful production of a transgenic animal will be affected by several events: the inserted DNA will incorporate into the chromosomes of only a percentage of the embryos developing from the microinjected eggs; the DNA will incorporate at different genetic locations; and different numbers of copies of the DNA will incorporate in different embryos. Therefore each embryo has the potential to become a unique transgenic mouse even though the same quantity and type of DNA was injected into genetically identical fertilized eggs. Not all manipulated, fertilized eggs become live-born transgenic mice. Losses occur at every step from injection through gestation and delivery (ARENA-OLAW, 2002).

Although an individual mouse may carry transgenes, it cannot transmit the transgene to its offspring, unless the cDNA incorporates into germ cells. A "founder" is a mouse that passes the transgene to its descendants. Thus, many fertilized eggs must be microinjected to obtain a few transgenic mice, and only a few of the transgenic mice will be founders of a particular transgenic line (ARENA-OLAW, 2002).

## Knockout and Knockin Mutants: Animals with Targeted Mutations

*Targeted mutation* refers to a process whereby a specific gene is made nonfunctional ("knockout") or, less frequently, made functional ("knockin"). A mouse with a targeted disruption, or knockout, of a specific gene is typically created through the embryonic-stem-cell method. This arduous method requires the occurrence of several low-probability events. First, the gene in question must be identified, targeted, and marked precisely. This has been accomplished for an astounding number of murine genes during the last several years (Harris and Ford, 2000; Takahashi et al., 1994). Second, mouse embryonic stem cells must be harvested and cultured. Third, a mutated form of the gene of interest is created (the mutation, or altered order of nucleotides, renders the gene inactive). Fourth, the mutated gene is introduced into the cultured stem cells by using microinjection or electroporation transfection (Tonegawa, 1994); a very small number of the altered genes will be incorporated into the DNA of the stem cells through recombination (Sedivy and Sharp, 1989). Fifth, the mutated embryonic stem cells are inserted into otherwise normal mouse embryos (blastocysts), which are then implanted into a surrogate mother (Boggs, 1990; Le Mouellic et al., 1990; Steeghs et al., 1995). All the descendant cells from the mutated stem cells will have the altered gene; the descendants of the original blastocyst cells will have normal genes. Thus, the newborn animals will have some cells that possess only a copy of the mutant gene and some cells that only possess the normal (wild-type) gene. This type of animal is called a chimera. If the mutated stem cells are incorporated into the germ line (the cells destined to become sperm or ova), some of the gametes will contain the mutant gene. If the chimera is bred with wild-type mice, some of the offspring will be heterozygous for the mutation (possess one copy of the mutant gene). If the heterozygous mice are interbred, about one-fourth of their offspring will be homozygous for the mutation. The homozygous mice become the founders and can be interbred to produce pure lines of mice with the gene of interest "knocked out" (Galli-Taliadoros et al., 1995). As a result, the product that the gene typically encodes will be missing from the progeny (Sedivy and Sharp, 1989).

For technical reasons, most of the stem cells used in targeted-gene deletion studies were derived from mice of the 129/SV strain. The 129/SV stem cells were typically implanted into C57BL/6 blastocysts (Soriano, 1995). The resulting "mixed" offspring are often backcrossed to the C57BL/6 (background) strain. After 10 backcrosses, the mutated strain is considered a congenic strain, identical with the C57BL/6 background strain except at the site of the altered gene.

There are several important advantages of using knockout mice: (1) disabling a gene often results in a precise and "clean" ablation, (2) the effects of the gene product can be abolished without the side effects of drugs, and (3) genetic manipulation may be the only way to determine the precise role of the gene product particularly in behavior. The use of new inducible knockouts, in which the timing and placement of the targeted gene disruption can be controlled, will refine and extend the usefulness of genetically modified animals in neuroscience and behavioral research.

One drawback in the use of knockout animals is lethal mutations. The products of many genes are essential for normal function, and inactivating a gene may prove lethal because of gross morphologic or physiologic abnormalities. For example, knockout mice with targeted disruption of either the parathyroid hormone-related peptide, β-1 integrin subunit, or β-glucocerebrosidase genes die in utero or immediately after birth (Karaplis et al., 1994; Stephens et al., 1995).

## Development of Animal Protocols Involving Genetically Modified Animals

### Disease Profiling

The first step in developing a protocol to produce or use genetically modified animals is to determine the disease profile that any particular animal or group of animals is likely to experience during the course of normal life or as a result of experimental use. Some genetically modified animals are created to develop a disease spontaneously, but others may develop a severe or debilitating disease even if the disease is not the intended outcome.

Genetically modified animals are used in a wide array of experimental studies. They can be used in studies of the pathogenesis and therapy of a primary disease, of a concurrent disease or associated clinical problems, or of aging and longevity. However, many of the animals will never be used in experimental studies but rather are maintained as breeders. The PI, IACUC, and veterinarian all need to develop a general health profile of a given strain that is relevant even to nonexperimental animals (breeders and animals intended for but not yet included in a study). The potential adverse effects of the genetic modification itself have to be considered.

For an established strain of genetically modified animal, the literature may provide a good description of the expected phenotype and the course of its development. However, the full repertoire of a gene's effects may not be envisioned, or a gene's functions may be unknown at the time a knockout or transgenic animal is created. That uncertainty can make the health-related consequences of developing a knockout or transgenic strain difficult to predict. Many modified mice are generated and maintained for purposes of discovery rather than hypothesis-testing. An example of the "discovery" approach is the use of random mutagenesis, which may create animals whose individual phenotypes are theoretically unpredictable. Because some of the new strains of mice may spontaneously develop problems that are painful or debilitating, assessment strategies and endpoints for these animals must be considered before their generation or their experimental use. Such information is typically solicited by the IACUC as part of the animal-use protocol evaluation.

### Animal-Number Estimates

When submitting an animal-use protocol to develop a genetically modified animal, neuroscientists must include an estimate of the number of animals to be used (not including experimental manipulations). Determining the number can be challenging because the process to develop a genetically modified animal is subject to unpredictable outcomes. A detailed presentation of a method for estimating the number of animals needed to develop a genetically modified founder mouse can be found in Appendix B.

After founder mice have been identified, 80–100 mice may be needed to maintain and characterize a line. That assumes that up to five breeder pairs per line are needed, that there is no unusual infertility, and that adequate numbers of weanlings are produced for genotypic and phenotypic characterization (ARENA-OLAW, 2002) Appendix B also contains extensive information on calculating animal estimates for colony breeding and experimental use.

Breeding of a congenic strain by using "speed congenics" requires a significant number of animals. Speed congenics is the process by which the DNA of each mutated animal is screened to select animals with the most genetic similarity to the background strain; this reduces the number of back-crosses necessary to develop the congenic strain. Usually, at least 750 mice are required, assuming a breeding colony of 10–12 breeding pairs and adequate progeny for phenotypic and genotypic characterization. If the homozygous mutant is infertile, the congenic strain must be developed by using intercross matings, and the number of mice needed is about 1,200 (ARENA-OLAW, 2002).

Because development and maintenance of genetically modified animal colonies require large numbers of animals, animals may be produced that are determined not to be useful for a particular project. Those animals may be useful for another project and should either be transferred to that project or culled from the colony.

## Animal-Welfare Considerations

The debility that genetically modified animals may experience is a cause of concern. It is important to provide as much support and comfort for mutant animals as possible. Some strains may require specific husbandry interventions to enable or promote well-being. For example, mice with targeted deletion of the gene for neuronal nitric oxide synthase (NOS-1 -/-) develop defects that model the clinical idiopathic voiding disorders that can affect to 10–15% of men and women. These mice have hypertrophic dilated bladders, dysfunctional urinary outlets, and increased urinary frequency (Burnett et al., 1997). They require extra bedding and more frequent cage changes than wild-type mice. Other examples of special husbandry interventions are those prone to audiogenic seizures, which must be housed in quiet environments, and those with ataxia or paralysis, which may require special provisions to enable access to food and water.

Close scrutiny of genetically modified animals during routine daily observation by the animal-care personnel may be warranted. Animal-care personnel often discover disabilities and abnormalities in genetically modified animals (such as motor deficits or anorexia) and should be trained to recognize them. Additional training of the animal-care staff to include practical information on the special needs and common problems associated with specific strains is recommended.

The dramatic growth in the use of genetically modified rodents, primarily mice, creates substantial challenges for timely and effective assessment of animal health and well-being. Many institutions house large populations of genetically modified mice with a wide array of deficits that affect physiologic homeostasis and behavior. The popularity of high-density, individually ventilated caging systems for housing these valuable mice adds barriers and challenges for effective observation and increases the importance of a careful and systematic examination of individual animals during scheduled cage-maintenance activities.

## General Health Assessment

The general health of novel genetically modified animals should be assessed soon after their availability and before the onset of complex behavioral analyses (Crawley, 1999). Identifying potential health problems early is critical to providing appropriate care. Undetected health problems can also skew the assessment of more complex behaviors—such as learning and memory, aggression, mating, and parenting, so it is essential to identify problems before behavioral phenotyping (see "Behavioral Screening of Genetically Modified Animals" in Chapter 9).

For mice, a general health assessment starts with a brief evaluation of body mass, core body temperature, and appearance of the pelage (fur). Neurologic reflexes should be assessed, including the righting reflex, the eye blink, and the ear and whisker twitch in response to tactile stimuli (Crawley, 1999). Any of the following symptoms should be recorded, treated if necessary, and considered when behavioral phenotyping is later conducted: self-mutilation, guarding, vocalization (with or without stimuli associated with pain), hunched posture, inactivity, lethargy, rough hair coat, no response to mild stimuli, increased heart or respiratory rate, anorexia for longer than 24 hours, weight loss greater than 20%, decrease in weight gain compared with aged-matched controls, and lesions (such as swelling, redness, and abnormal discharges). Any obvious deviations from the typical naturally occurring behaviors (ethogram) of mice should be noted. The mouse ethogram includes such behaviors as sleeping, resting, locomotion, grooming, ingestion of food and water, nest-building, exploration, foraging, and fear, anxiety, and defensive behavior (Brown et al., 2000).

After an initial health assessment, daily observation of the genetically modified animal should include an assessment of general activity levels, posture, haircoat condition, the presence of scratching or self-mutilation, and the general condition of the cage. When the cage is manipulated, as during cleaning, animals can be more closely examined for additional characteristics, such as response to handling; unexpected vocalization; ulceration; masses; abnormalities of the eyes, ears, nose, and mouth; palpable hyperthermia or hypothermia; and general body condition.

### Behavioral Assessment

Subsequent to the general health assessment, sensory and motor testing should be carried out followed by behavioral testing (e.g., anxiety behaviors). Behavioral assessment should proceed as soon as sufficient numbers of transgenic animals are available to identify sensory, motor, or motivational deficits that may compromise the well-being of the animals. Behavioral screening is discussed at length in Chapter 9, "Behavioral Screening of Genetically Modified Animals."

## Pain, Distress, and Endpoints

The elimination of all pain and distress from all affected animals is unlikely, inasmuch as the diseases being modeled in genetically modified animals are often associated with pain or distress that cannot be relieved in human patients. Achieving a balance between animal well-being and research objectives is essential to obtaining valid answers to questions about the causes, treatment, and preventions of such diseases in humans.

When a neuroscientist initiates assessment of a new genetically modified animal, information about clinical abnormalities associated with the phenotype and special husbandry requirements usually are not available. The investigator must, however, include general humane endpoints in case a severe debilitating phenotype develops and should provide the IACUC with this information in writing when the new mutant has been developed or at the next annual review of the animal-use protocol.

When a genetically modified animal strain has been characterized, the standard of "normal" for a mutant animal may or may not be the same as that for a nonmutant animal (ARENA-OLAW, 2002); therefore, clinical signs that would be used as humane endpoints in normal animals may be inappropriate as endpoints in genetically modified animals. If the mutant phenotype does not affect the general welfare or clinical well-being of an animal, the same standard of "normal" may be used for mutant and nonmutant animals. In the case of mutants whose phenotype involves clinical abnormalities, the standard for "normal" may have to be modified to include the expected phenotype. For example, 8-month-old mice lacking the gene for a key enzyme that encodes ganglioside biosynthesis (GM2/GD2 synthase) develop substantial neuropathologies, motor incoordination and an abnormal gait (Chiavegatto et al., 2000). As these mice age, muscular weakness progresses, and the standard of "normal" for GM2/GD2 synthase knockout mice includes difficulties in locomotion, which in a nongenetically modified animal might be one criterion of a humane endpoint.

Humane endpoints for mutant animals should be established on the basis of the ability of the mutant to access and consume food and water, the response of the mutant to stimuli, and the general condition of the mutant (for example, it is excessively underweight, it shows progressive weight loss, it doesn't groom, it has a hunched posture, or it has sensorimotor deficits).

The specific use of a genetically modified animal will influence the type of endpoint that is described in the animal-use protocol and the circumstances in which an endpoint decision will be implemented. For example, an animal that develops a clinical problem while in a study of the prevention of disease development could potentially be euthanized earlier than one involved in a study of disease therapy. A nonexperimental animal (a breeder or an animal intended for but not yet part of a study) that develops a substantial clinical problem should be euthanized. A maximum holding period should be set to avoid the development of predictable problems in strains of mice that have debilitating phenotypes.

Endpoint issues generally apply to the entire life of genetically modified animals. Therefore, endpoints become relevant both in the context of experimental procedures and with regard to the potential pain or distress that is caused by the genetic modification itself. Care must be taken to provide general endpoints in the animal-use protocol for the period in which the initial colony is being developed and the phenotype of the animals is first characterized, as well as for experimental and nonexperimental animals.

# PART II

## Applications to Common Research Themes in Neuroscience and Behavioral Research

# 4

# Survival Studies

## ANATOMIC STUDIES

Anatomic studies are used to evaluate the nervous system by examining the cellular organization or chemical composition of specific brain regions or by examining how different brain regions are related by afferent or efferent connections. These studies most commonly involve either the use of tracer substances to label and visualize neural pathways or the use of lesion techniques to destroy a discrete area of brain cells and examine the course of degenerating fibers. Electrolytic and radio frequency techniques, as well as those using neurotoxins, can be used to make brain lesions. Stereotaxic approaches are often used to make more focal lesions or lesions in deeper brain structures. When tracers are used, they are injected into the nervous tissue, where they are incorporated into neuronal cell bodies and/or processes and then transported anterograde or retrograde. Transport of tracers and degeneration of fiber pathways generally occur over a period of several days after injection or a lesion; therefore, an animal must be allowed to survive for a short period before being sacrificed for study of its nervous system. The possibility of using labeled substances, such as manganese, in combination with brain imaging to trace anatomic connections is also developing (Saleem et al., 2002); the same animal can be examined repeatedly with this technology, so it reduces the number of animals needed for a particular study (see "Imaging Studies," below).

Various factors will determine whether and to what extent the IACUC and the investigator need to exercise flexibility in interpreting and implementing the recommendations of the *Guide*. Those factors include the invasiveness of the procedures, the surgical setting, whether multiple injections are necessary, and the characteristics of the injected substance. This brief discussion is derived from page 12 of the NIH workshop report, *Preparation and Maintenance of Higher Mammals During Neuroscience Experiments* (NIH, 1991).

### Grading the Invasiveness of the Procedure(s)

> The invasiveness of the procedure required to inject a tracer will establish whether it constitutes a major survival surgery, which should be performed in a facility intended for that purpose. Thus, while an injection into the eye is a relatively minor surgical procedure similar to a biopsy, an injection into a central brain structure, which usually requires performing both a craniotomy and a durotomy, is usually considered a major surgical procedure (NIH, 1991).

For further discussion of major versus minor surgery, see "Surgery and Procedures" in Chapter 3.

### Modified Surgical Settings

Anatomic studies are often performed in modified surgical settings (for a discussion of the characteristics of a modified surgical setting, see "Surgery and Procedures" in Chapter 3). The reason for allowing an exception to the *Guide's* recommendations about performing major survival surgery in a dedicated surgical facility is that it may not be possible to sterilize the necessary experimental equipment (amplifiers, oscilloscopes, audio monitors, micropipette pullers, pressure microinjection devices, and micromanipulators) or move it into the dedicated surgical facility (NIH, 1991). Other factors that may influence an IACUC's decision to approve a modified surgical setting are (NIH, 1991):

> (1) the use of a radioactive tracer substance;
> (2) the need to manufacture, fill, and position into the brain several injection micropipettes during the course of a single procedure;
> (3) the relatively short duration of the post-surgical survival period (2–4 days);
> (4) the suitability of the proposed laboratory area for aseptic surgery;
> (5) the infrequency of the procedure (less than once per month, on average);
> (6) verification of the absence of post-surgical infection or other complications in a series of animals from a pilot project.

### Multiple Injections

> The scientific requirements of certain experiments may require subjecting a single animal to multiple injection procedures. As a result, when the injection site is a central brain structure, it will be necessary to subject an animal to multiple major survival surgeries. Experiments of this sort include those in which two or more tracers are to be injected and it is known that they require

> markedly different survival periods for transport to occur. Another example is experiments in which injections will be made at different ages in the same animal in order to label the arrangement of connections at different stages in the development of a neural pathway (NIH, 1991).

Each of those situations would require an IACUC exception to guidelines and regulations on multiple major survival surgeries. In both examples, the surgeries are "related components of a research project" (NRC, 1996), and IACUCs may choose to grant approval for these multiple major survival surgeries.

### Aseptic Technique

> Standard aseptic technique is used to expose the surface of the occipital cerebral cortex. At this point the animal often is redraped and the surgeon breaks sterility in order to fill and insert the injection micropipettes into the micromanipulator, position the pipettes into the brain, and adjust and activate the pressure injection device. Whether or not the micropipettes are able to be sterilized prior to surgery will depend on several factors, including the material out of which the pipette is made and the substance being injected. . . . Throughout this period, the surgeon has no direct contact with the wound site or the surgical field. When the injections have been made, the surgeon re-gloves (and re-gowns, if necessary), the top level of drapes is removed, and the wound is closed using standard aseptic technique (NIH, 1991).

The topic of micropipette sterility is similar to the discussion of implant sterility in the succeeding section ("Asepsis and the Introduction of Devices or Implants into Neural Tissue").

### Characteristics of the Injected Substances

> Another important consideration is the chemical properties of the substance to be injected. For instance, many tracers are sensitive to temperature and cannot be heat sterilized prior to injection. Others may exhibit high levels of tissue toxicity and can cause a marked local inflammatory response at the injection site. Some neural tracers are radioactive and proposed procedures for their use must be reviewed for compliance with institutional policy on radiation safety (NIH, 1991).

## NEUROPHYSIOLOGY STUDIES

### Neurophysiology Experiments in Awake, Behaving Animals

Neurophysiology experiments in awake, behaving animals have fundamentally shaped our understanding of the processing of information throughout the brain because they provide the most direct information about neural signals.

Studies linking neurophysiology and brain imaging are useful for clarifying the meaning of the brain activation seen during functional-imaging experiments. The behavioral repertoires of many mammals resemble those of humans, and data generated using awake, behaving animals can have considerable relevance when extrapolated to humans. Awake, behaving animals make it possible to study the "higher" functions of the brain by involving the active participation of the animal. Trained animals not only can serve as subjects in experiments on motor control, but also can be sophisticated participants in psychophysical studies of the processes involved in perception and memory (NIH, 1991). In addition, the results of brain-recording experiments have probably had more influence on the fields cognate to neuroscience than any other kind of experiment; their influence is evident in fields as disparate as behavioral psychology, image-processing, and computer design.

Experiments on an awake, behaving animal generally involve a long initial phase during which the animal is trained to perform a task. Once the animal is trained, experimental sessions are held several days a week for months and sometimes years, during which specific physiologic variables are measured. For example, a nonhuman primate may be trained to look at a particular object on cue. Then, during subsequent experimental sessions, the researcher measures the neural activity associated with the eye movement. Monitoring a physiologic process, whether it is the experimental variable (the electrical activity in the brain) or the acquisition of the behavioral task (the eye movement), frequently necessitates the implantation of various devices. In the previous example, eye coils may be implanted to monitor eye position and microelectrodes may be placed in the brain to measure neural activity.

Neurophysiology experiments on an awake, behaving animal frequently require that the animal be confined to a restricted working space for many reasons, such as to enable precise positioning of recording or stimulating electrodes into the correct region, to stabilize the spine or head for neurophysiologic recording, to maintain the animal's posture in relation to the behavioral task, to maintain the orientation of the animal relative to a sensory stimulus, to restrict the range of reach of an animal to prevent it from accidentally damaging implanted equipment that is exposed during recording sessions (e.g., electrodes or lead wires), or to prevent movement that would induce errors or variation in the experimental data (Lemon, 1984c). Restraint of an awake, behaving animal often involves transferring the animal to a special apparatus, such as a restraining chair or an operant chamber. There is a long tradition of studying the neurophysiology and behavior of rodents in various kinds of mazes (including water mazes), running wheels, or open-field areas (Porsolt et al., 1993). Depending on the experiment, the apparatus into which an animal is placed may or may not be inside a larger chamber that is designed to attenuate extraneous visual or auditory stimuli during the experimental session (Ator, 1991). Whatever specialized chamber is used, the animal remains in it for the duration of the experimental session.

Some types of neurophysiology experiments require that a probe be placed in the brain only during the actual experimental sessions, either to stimulate the brain (stimulating electrode), to record electrical activity (recording electrode), or to sample the fluid in the interstitial space (microdialysis probe). Owing to their fragility, or the need to reposition them to sample from a different area of the brain, these probes are removed at the end of each recording session. The use of these types of devices requires the implantation of chronically indwelling hardware called guide cannulae or chambers. When these are implanted, a piece of the skull is removed (craniotomy) and the hardware is placed over the hole and attached to the skull. The hardware is hollow, allowing free access to the brain, and is filled with a sterile solution (typically saline) and capped with a sterile cover to prevent introduction of microorganisms. When the investigator needs to place a probe to begin an experimental session, the cover to the guide cannula or chamber is removed and the probe is introduced into the brain.

Major surgery to implant hardware devices for head restraint, data collection, and stimulation can be accomplished with standard aseptic surgical techniques and typically can be performed in a facility dedicated to aseptic surgery (see Gardiner and Toth, 1999, and Lemon, 1984b, for discussions of surgical issues related to cranial implants). When implanting guide cannulae or chambers, the size of the craniotomy should be large enough to allow access to the structure being studied without unnecessarily exposing neural tissues (Lemon, 1984a). If an implanted device is necessary during the training of the animal, the animal should be conditioned to the training environment prior to any surgery. In this way, animals that will not accept training can be removed from the study before they are subjected to an unnecessary surgery.

## Neurophysiology Experiments in Anesthetized Animals

Cases where recordings are made while an animal is anesthetized raise critical questions regarding anesthesia, maintenance of physiologic status, and monitoring of the animal's condition. The choice of anesthetic must satisfy the need of the experimenter to perturb neuronal status as little as possible while ensuring that the animal remains free of pain and distress. Maintaining an anesthetized (and sometimes immobilized) animal in appropriate physiologic condition is a considerable technical challenge (see "Prolonged Nonsurvival Studies" in Chapter 5). Monitoring both the anesthesia and the animal's general condition requires careful attention to a number of measures. Although animals in some studies are used in repeated experiments (with intervening recovery periods), in other cases they are maintained under anesthesia for long periods of time for nonsurvival studies (see Chapter 5). If there will be repeated sessions of prolonged anesthesia, special attention should be paid to maintaining the animal's normal physiologic status between anesthetic sessions.

## Animal Care and Use Concerns Associated with Neurophysiology Experiments in Awake, Behaving Animals

Many of the animal care and use concerns associated with recording in awake, behaving animals were discussed in Chapter 2 and 3. They include the use of food and fluid regulation, monitoring animals for signs of pain or distress, and the general use of restraint. Additional animal welfare issues specific to or frequently encountered in neurophysiology experiments include head restraint systems, chairing nonhuman primates, multiple survival surgeries, modified surgical settings, asepsis during introduction of probes into the brain, monitoring the site surrounding implanted devices or hardware for signs of infection, dealing with rejected or failed implants, maintaining chambers free of infection, and periodic durotomy.

## Restraint During Neurophysiology Experiments in Awake, Behaving Animals

Most experiments that involve the monitoring of neural activity require some limitation of the animal's working space and/or freedom of body movement (NIH, 1991). Many forms of restraint are acceptable as long as the particular procedures for accomplishing and monitoring restraint are well justified and consistent with the *Guide*, and the period of restraint is as short as possible. Animals that are restrained must be monitored closely to ensure that the restraint method permits reasonable postural adjustment, does not interfere with respiration and does not cause skin abrasions or bruising. If ulceration or bruising develops, the animal should be removed from the study until the injured area is fully healed, and adjustments should be made to correct the source of the problem.

In some neurophysiology studies, the restraint is the independent variable in an experiment (for example, to study the physiological responses believed to be affected by unfamiliar restraint). However, in most cases, the restraint is not a variable in the experiment, and a training phase is carried out to habituate the animal to the restraint before the experiment begins. Because animals in behavioral experiments are handled frequently (often 5 or even 7 days a week), they usually become habituated to the head restraint, tether, or chairing quickly. The best evidence of behavioral adaptation to restraint is voluntary movement into the device (NIH, 2002) and performance of the behavioral task once there.

In experiments in which animals are tethered to treadmills or other devices used to study locomotor behaviors, care should be taken to ensure that they cannot become trapped in the apparatus (see " Exercise" in Chapter 8). That may require intensive and continuous monitoring of animals during training and recording sessions. As mentioned before, appropriate and thorough habituation to the apparatus before experiments begin can substantially reduce the risk of entrapment and the distress that could arise with the use of the device.

### Tethering

In some types of neurophysiology experiments, the animal's activity may be restrained with a tether. For example, in experiments involving intravenous drug self-injection or intragastric drug delivery (Lukas et al., 1982) the animal may have a chronic indwelling intravenous or intragastric catheter that (Lukas and Moreton, 1979; Meisch and Lemaire, 1993) exits from a site on the back (typical in monkeys) or the top of the head (typical in rats and cats), and is threaded through a protective tether that is connected to a swivel. The tubing emerges from the swivel and is connected to a pump, which is used to deliver the drug. Monkeys that have been fitted with chronic indwelling catheters often wear specially designed vests, shirts, or harnesses to protect the catheter exit site. Habituation of an animal to a harness-tether arrangement is best carried out well in advance of the planned date of implantation of the catheter. Inspection of the animal during the habituation process allows the experimenter to determine whether the vest or tether fits well and permits adjustments to prevent discomfort.

### Head-Restraint Systems

Head-restraint systems minimize the movement of the head during neurophysiology experiments without causing discomfort if the animal is properly conditioned (NIH, 2002). Hardware, generically called a head-holder, is implanted chronically on the animal's skull. Three different styles of head-holders are generally used: implantable, halo, and headpiece (Lemon, 1984a). Small screws or bolts and dental acrylic or bone cement anchors the head-holder to the skull. Then, during a training or experimental session, the head-holder is attached to a freestanding platform to immobilize the head. Besides minimizing movement, these systems provide a structural element to which to anchor connectors from other surgically implanted monitoring devices, such as eye coil wires, chronically implanted recording electrodes, or indwelling cannulae for delivery of pharmacological agents. They also can provide a superstructure through which microelectrodes are introduced into the brain for the recording of neural activity. Animals should be properly conditioned to the restraint to eliminate any discomfort or stress that might be associated with it (see "Physical Restraint" in Chapter 3).

### Chairing Nonhuman Primates

Macaques and squirrel monkeys can be trained to move voluntarily from the home cage into a restraint chair (Ator, 1991). Commonly, nonhuman primates wear either a collar with a small metal ring attached or a collar with a slot to which the pole directly attaches. The monkeys acclimate to having a chain clipped to their collar; the chain is then pulled through a ring at the top of a metal pole.

Squirrel monkeys usually grasp the pole and ride to the chair on it, while larger monkeys, such as adult macaques, learn to walk to the chair. By holding one end of the pole snugly at the collar and pulling the chain down to the other end, the experimenter can control the monkey's movements and proximity and thus be protected from the possibility of a bite in the process of training and transfer. Larger monkeys can be trained to move from the home cage into a smaller shuttle device that can be wheeled to the experimental chamber. Treats may be used during the various steps of training the monkey to cooperate in the transfer process and sitting in a chair. While the amount of time that an animal is chaired can be gradually extended during the training process, the animal should not live in the chair. In instances where long-term (greater than 12 hours) restraint is required, the nonhuman primate must be provided the opportunity daily for unrestrained activity for at least one continuous hour during the period of restraint, unless continuous restraint is justified for scientific reasons and approved by the IACUC (AWR 3.81 (d)).

## Multiple Survival Surgeries

In many experiments using awake, behaving animals, the implantation and maintenance of recording devices, head restraint devices, and stimulation devices necessitates multiple major survival surgeries. The use of multiple surgeries in these experiments, including surgeries to repair implants, is permitted by the *Guide* because they are related components of a research project, they will conserve scarce animal resources, or they are needed for clinical reasons.

The need for multiple major surgeries may arise for several reasons. In some cases, it arises for clinical reasons. For example, a head restraint device and a chamber may need to be implanted in close proximity. Implanting them during one long surgery could undermine the structural integrity of the skull, create an extremely large wound, and increase the risk of infection. Multiple major surgeries may also be necessary due to limitations of the experimental devices used. For instance, eye coils typically will function reliably for a limited period of time after implantation. If a prolonged training period with head restraint is necessary before experimental sessions can begin to monitor eye position using eye coils, implanting the head restraint device and the eye coils during one surgery prior to the start of training could mean the eye coils will not function reliably during the subsequent experimental sessions. In this case, performing a second major survival surgery is necessary and justified to ensure that the eye coils will function reliably during the experimental session.

Performing multiple major surgeries may also be the best surgical approach if doing so allows a major surgery to be performed in an aseptic surgical suite, rather than in a modified surgical setting. Often, probes must be positioned precisely in the brain of an awake, behaving animal by means of a head restraint device or recording chamber that is implanted on the skull in stereotaxic coordinates. However, it may be impossible to move all the equipment necessary both to implant the device or chamber on the skull stereotaxically and to monitor the output of the neurophysiological probes into a suite dedicated to aseptic surgery. Rather than performing a single long surgery to implant the head restraint device and then position the implanted probes in a modified surgical setting, performing multiple surgeries may be preferable. In this way, the head restraint device can be implanted in the aseptic surgical suite and the animal can recover from the surgery and heal. Then a second smaller surgery to place the electrodes could be performed in the modified surgical setting. This minimizes the potential for infections and subjects the animal to two short surgeries rather than one prolonged surgery.

Multiple major surgeries may also be required to maintain the viability of implanted devices. Though all percutaneous implanted devices are designed so that the skin can heal around them and the devices can be used without causing the animal pain or distress, it may be necessary to replace electrodes or eye coils that no longer function or to replace implanted hardware that has failed or been rejected.

In all of these cases, PIs, veterinarians, and IACUCs must work together to balance the animal's well-being and the scientific goals of the experiment. Consideration of such factors as the use of scarce or conserved species and the disposition of individual animals (especially the case with higher mammals such as nonhuman primates) will influence the decision of how many survival surgeries are acceptable.

There is no need to treat all procedures for the clinical management of an implanted animal as major survival surgeries that must be performed in a facility dedicated to aseptic surgery. Procedures, such as treating surgical wounds as they heal, cleaning and maintaining implanted devices, and removing the granulation tissue that typically forms over the dura mater inside chronically implanted recording chambers (Lemon, 1984a; Toth and Gardiner, 1999), are commonly performed under light anesthesia in a laboratory setting, using aseptic techniques within a local sterile field (NIH, 1991). Classifying those procedures as major or minor surgeries according to regulatory guidelines is not straightforward (see "Surgery and Procedures" in Chapter 3), demanding that professional judgment, guided by outcome or performance-based consideration, be employed. Certainly, the long tenure of these animals in the research setting and the many hours devoted to their training militates in favor of exercising maximum precautions to avoid infection. However, the majority of these procedures are brief and innocuous, with minimal risk of infection, and the investigator, veterinarian, and IACUC should use professional judgment to balance the well-being of the animal with the practicality of performing the procedure in a facility dedicated to aseptic surgery.

## Modified Surgical Settings

Sometimes, it is necessary to implant recording or stimulating devices using neurophysiologic responses to identify the correct location in the brain. This surgery should be performed in a facility dedicated to aseptic surgery whenever possible. However, if the procedure requires specialized equipment that cannot be sterilized or moved into a dedicated surgical facility, then all or a portion of the surgery may be performed in an approved modified laboratory setting. In these cases, the surgery sometimes can be performed in two steps. The first step—which often entails the implantation of the hardware for head restraint and opening of the skull—is performed in a dedicated aseptic surgical facility. A temporary cap is placed over the opening of the skull. On the following day, the animal is taken to the laboratory and restrained with the hardware that has been implanted for head restraint, and the temporary cap is removed. A microelectrode, micropipette, or microdialysis probe may then be implanted into the brain, maintaining asepsis in the area immediately around the site and using specialized equipment to position the device accurately. If the two-step approach is not feasible and if the laboratory can be sanitized and prepared to allow aseptic technique, the entire procedure may be performed as a single survival procedure in the laboratory (for more discussion of this subject see "Surgery and Procedures" in Chapter 3).

## Animal Care and Use Concerns Associated with Introduction of Probes into Neural Tissue

Questions about sterility arise when considering the implantation of probes, such as microelectrode, micropipette, and microdialysis devices, into neural tissue. Most implanted probes can be sterilized, but this may not always be the case for sensitive or delicate probes such as microelectrodes or micropipettes, as there is no consensus on whether they can be sterilized without degrading their performance. Many laboratories do not sterilize microelectrodes and micropipettes because of their fragility, and this practice does not seem to introduce infections into the brain. Currently, there is no published, systematic evidence that the use of micropipettes or microelectrodes that have not undergone rigorous sterilization before implantation has a deleterious outcome on experiments, on the brain, or on animal health. This could be because the materials and fabrication methods used to produce microelectrodes and micropipettes may result in their being relatively free of microorganisms without additional intervention. With this in mind, any material that will be inserted into or implanted in the brain should always be handled and stored with care to protect against contaminants. The above notwithstanding, whenever possible, probes should be sterilized or alternatively disinfected before they are inserted into neural tissue.

The success with which a probe can be sterilized or disinfected immediately before its use will depend upon several factors, including the materials out of which it is made. Existing options for *sterilization* include heat or gas methods, soaking in bactericidal solutions, and irradiation with ultraviolet light (Lemon, 1984a). In many cases, the materials used to manufacture probes may not withstand those rigorous sterilization procedures. In such situations, a method of *disinfection* should be used if possible, such as soaking in povidone iodine, chlorohexadine, or aqueous alcohols and then rinsing with sterile saline prior to insertion. If none of those options preserves the viability of the probe, attention to maintaining its cleanliness during handling and storage becomes even more important.

Investigators, veterinarians, and IACUCs should monitor for deleterious effects caused by nonsterile probes by developing performance-based standards for the histopathologic analysis of postmortem tissue specimens. As new methods become available to sterilize microelectrodes and micropipettes without compromising their utility, such as vaporized hydrogen peroxide, they should be implemented.

In some types of neurophysiologic experiments, probes such as microelectrodes, micropipettes, or microdialysis probes are introduced into the brain through guide cannulae or chambers at the beginning of the daily experimental session and removed at the end of the session. These types of probes are usually introduced without anesthesia, and their introduction typically does not require they be performed in a dedicated surgical facility, though aseptic technique when handling and inserting the probes is necessary to prevent infection. The brain itself lacks sensory endings, so the passage of these probes gives rise to no sensation. The dura mater does contain nociceptive fibers, primarily adjacent to large blood vessels (e.g. the middle meningeal artery) (Baker et al., 1999; Wolff, 1963); however, the insertion of probes through the dura mater usually evokes no reaction from an animal. On occasion though, an indication of momentary or minor discomfort may be noted. The U.S Government Principles state that "procedures that may cause *more than* (emphasis added) momentary or slight pain or distress should be performed with appropriate sedation, analgesia, or anesthesia." Accordingly, in most instances, the placement of probes in awake, behaving animals may be performed safely and humanely without sedation, analgesia, or anesthesia.

Potential adverse consequences of insertion of probes into neural tissue are infection or brain injury as a result of cerebral edema or hemorrhage. The likelihood of those deleterious effects is affected by the frequency of probe insertion, the location of the probe insertion site, the depth of penetration, the physical characteristics of the probe, the expected duration of experimental sessions, and the course of the experiment for each animal. Training laboratory personnel in identifying adverse reactions and fostering a team approach that includes veterinarians and husbandry staff will help to ensure the well-being of animals used in these types of studies.

### Monitoring the Site Surrounding an Implanted Device

Sites surrounding implanted devices or hardware, such as chambers, head-restraint devices, eye coils, nerve cuffs, electromyography (EMG) electrodes, etc., should be examined regularly for signs of irritation, infection, or device damage. Specifically, investigators and animal-care staff should watch for signs of inflammation or infection of the eye coils, along the subcutaneous length of eye coil leads, and near the sites where wires, chambers or other hardware devices are externalized. Similarly, the attachments for nerve cuffs around nerves or of EMG electrodes onto muscles should be closely monitored for signs of inflammation or infection. Leads from these types of implants often are externalized to connectors that are attached to the skin or bone. These connectors should be positioned so that they are not easily manipulated or broken by the animal. Implant protection may also necessitate the use of connector hoods or fitted jackets for the animals to protect the externalized wires or connectors. Like eye coil leads, the wires from other devices should be examined throughout their subcutaneous lengths and at the skin margins for any signs of inflammation or infection.

## Implant Failures

Unambiguous experimental endpoints should be established before any devices or hardware are implanted. These endpoints should indicate when devices or hardware should be removed because of failure, infection, or inflammation. Successful reimplantation after implant failure may be possible in some circumstances. Therefore, the necessary conditions for reimplantation of previously used or replacement hardware should be described in the animal-use protocol and approved by the IACUC. Anticipating the potential consequences of implant failure before its occurrence is crucial for the viability of the study and animal well-being. A team approach involving veterinary staff, caretakers, neuroscientists, and technicians is critical to the long-term success of experiments that use animals with chronic implants.

## Occupational Health and Safety

It is prudent to reiterate that risks, such as exposure to B virus, are associated with working with awake, behaving nonhuman primates (see "Experimental Hazards" in Chapter 2). Investigators, their laboratory personnel, veterinarians, and veterinary-care support staff should all be aware of the resources that provide information about appropriate precautions in these types of experimental settings. Investigators should make certain that their research personnel are fully trained in the proper handling, husbandry, and maintenance of nonhuman primates and, if necessary, in the disposal of devices and other materials that have been in contact with their tissues or fluids. To minimize the risk of personnel exposure to biologic agents or puncture, used probes should be disposed of in approved biological hazard sharps containers.

## IMAGING STUDIES

Developments in imaging technologies have led to groundbreaking advances in our understanding of neural and physiologic functions in normal and diseased humans and animals by offering a view of the living brain at work (Hoehn et al., 2001). The technologies are generally less invasive than other investigative scientific methods and offer an opportunity to address questions of structure and function without significant consequences for the research subjects (Balaban and Hampshire, 2001).

### Imaging Techniques

Several imaging techniques are used in animals. They include positron-emission tomography (PET), single-photon emission computed tomography (SPECT), magnetic resonance imaging (MRI) and functional MRI (fMRI), nuclear magnetic resonance imaging or spectroscopy (NMR), near-infrared spectroscopy, ultrasonography, computed tomography (CT) and optical imaging (Balaban and Hampshire, 2001; Hoehn et al., 2001; Rolfe, 2000). Some of the techniques, such as PET and SPECT, enable measurement of blood flow, oxygen and glucose metabolism, receptor density, or drug concentrations in regions of the living brain (Mathias, 1996). Others, such as MRI and NMR, provide imaging of superficial and deep brain structures with a high degree of anatomic detail. High-field MRI, SPECT, and PET techniques can also be used to provide in vivo longitudinal evaluation of receptor binding and gene expression following gene therapy (Auricchio et al., 2003; Kasper et al., 2002).

Each of those techniques allows researchers to test hypotheses about the functions of different regions of the brain on the basis of functional composition or physiologic activity. The hypotheses can often be explored further with human subjects performing specific tasks during PET, SPECT, or fMRI. However, many of the technologies provide even better resolution when used in small mammals, providing more information about physiologic function than can be obtained with human subjects (Balaban and Hampshire, 2001). Animal models enable variables associated with specific diseases to be manipulated and controlled to a degree that is not possible with human patients. Furthermore, individual animals can be evaluated repeatedly during the course of a disease or can serve as their own control instead of sacrificing large groups of animals at different time points, and thereby reducing the number of animals used (Hoehn et al., 2001).

### Animal Preparation and Maintenance During Imaging Studies

Imaging generally requires anesthesia so that the animal remains motionless throughout the duration of image collection. The exception is ultrasonographic images, which can be collected from a restrained nonanesthetized animal, provided that the process does not create substantial stress in the animal. Conditioning animals to the type of handling associated with the scans obviates anesthesia. In fact, PET and fMRI scanning has been conducted on conscious monkeys that have been trained to sit in a chair (Stefanacci et al., 1998; Tsukada et al., 2000). However, it can take much time and effort to train the animals (Tsukada et al., 2000).

Generally, animals are sedated or anesthetized and then intubated either before or after transportation to the imaging facility. Because imaging facilities are rarely close to the vivarium, the methods by which animals will be transported to and from the imaging site must be considered when animal-use protocols involving these techniques are being prepared. Special attention must be given to the unusual occupational health and safety risks associated with transportation, including exposure of the transportation route or the imaging facilities to animal tissues or fluids; training and supervision of research and imaging personnel; and development of procedures for dealing with emergencies that arise during imaging or transport (such as bites and scratches). Furthermore, the personnel and methods used to monitor the animals and to administer appropriate care to ensure their well-being during imaging should be identified. Often, animals are imaged after normal business hours using facilities primarily dedicated to humans (such as at hospitals). The imaging facility professional staff may not be onsite after business hours to assist if there is a problem with the equipment, so identifying a member of the professional staff to contact in the event of an emergency may be necessary.

### Special Considerations of Animal Maintenance in the Imaging Environment

Some of the features of imaging machines that make them powerful tools create an environment that may be inhospitable to routine maintenance of anesthesia and monitoring of animals. For instance, the strong magnetic field associated with an MRI machine may damage ferromagnetic components in monitoring devices or traditional ventilators (Chatham and Blackband, 2001; Kanal et al., 2002) and indeed may actually attract ferromagnetic devices or standard surgical equipment to the magnetic-field coil. This can result in injury to personnel assisting in scanning or to experimental animals, and may damage the monitoring device and scanner (Chatham and Blackband, 2001). Before an animal is imaged with any device that creates a strong magnetic field, the research staff must ascertain that the animal does not have any ferrous implants. A variety of implants, made of nonferrous materials, are available and are suitable for use with imaging equipment.

Monitoring equipment that is compatible with the imaging equipment is available at imaging facilities and may be appropriate for monitoring animals. MRI-compatible physiologic monitoring capacity includes heart and respiratory rate, pulse oximetry, and temperature. Identifying the types of monitoring equipment that are available at the scanning facility and ensuring that it can be used with animal subjects are important considerations for these types of protocols. The ability to monitor the physiologic status of an animal during scanning is extremely valuable because direct observation and access to the animal may be reduced during image acquisition.

Maintaining an animal's body temperature during transportation to and from the imaging facility and during scanning improves the maintenance of anesthesia. Warming blankets often have metallic components or require a power source, so the use of portable, nonmetallic warming devices is advisable. These devices produce heat as a result of a chemical reaction or after microwaving. Covering the animal with blankets and using one of these warming devices is an effective way to maintain a favorable body temperature during relatively short imaging sessions.

As many imaging facilities are utilized both for human and animal scanning, the potential for cross-contamination exists. Human B virus exposure is always a concern when macaques are involved (Cohen et al., 2002) and human allergies to rodents, dogs, and cats are common (Wolfle and Bush, 2001). In addition, some animals may be susceptible to zoonotic diseases from humans; for example Old World nonhuman primates, such as rhesus macaques, are particularly susceptible to tuberculosis (Aiello, 1998b). Therefore, thorough disinfection of the equipment before and after its use may be warranted, especially when nonhuman primates are involved.

Finally, in positioning an animal in the scanner, care should be taken to maintain airway patency. Animals are usually intubated with an endotracheal tube during scanning, and this helps to ensure that the airway is not obstructed. Care should be exercised to prevent occlusion of the endotracheal tube and to prevent it from being dislodged during positioning. On completion of the imaging procedure, the animal may be extubated once a gag reflex and the ability to swallow are regained. The intravenous line should be removed, and the animal should be observed as it recovers from anesthesia before it is returned to its home cage.

## Occupational-Health Issues

The use of radioactive materials in imaging studies (such as in PET and SPECT imaging) poses specific occupational-health risks that should be considered as part of protocol development. Laboratory staff should be trained in the proper handling and disposal of radioactive materials. Furthermore, the potential for exposure to radiation from the animal and its bodily excretions after injection of radioactive tracers may have to be evaluated and appropriate actions taken to minimize the associated human health risks. Other considerations include thorough disinfection of the equipment if it is also to be used with human subjects or patients; the potential, during transportation of an animal to and from the imaging facility, for exposure of people who are not involved in the study; and determining whether the air exhausted from the imaging facility is recycled into other building areas.

## STEM CELL AND GENE-THERAPY STUDIES

Gene therapy is a technique involving the transfer of genetic material to an individual animal. Transfer can occur directly by administration of a foreign gene to an animal (in vivo) or indirectly through the introduction of genetically modified cells that contain a foreign gene (ex vivo) (NIH, 1995).

During in vivo gene therapy, foreign genes are introduced by administering DNA (naked or complexed with liposomes or proteins) (Cristiano, 2002; Lu et al., 2003; Templeton, 2002), RNA viruses (Quinonez and Sutton, 2002), or DNA viruses (Burton et al., 2002; Lai et al., 2002). To target the nervous system, the virus or DNA can be administered by microinjection into a specific region of the nervous system or by infusion into the bloodstream. Host cells are infected by the virus or will take up the DNA containing the foreign gene. The foreign gene will then exist in the host cells either episomally or integrated into a chromosome. The foreign gene may be chosen because it codes a desired protein, an antisense RNA (Sazani et al., 2002), or a potentially toxic protein (Dilber and Gahrton, 2001). The host cells will then express the foreign gene, changing the genetic profile of the host cells (NIH, 1995).

During ex vivo gene transfer, cells, such as fibroblasts, are removed from the body and genetically modified, often with the same methods used for in vivo gene therapy. The modified cells are then placed in a host animal (Murray et al., 2002).

Stem cell therapy is very similar to ex vivo gene transfer, except that the stem cell is the therapy, rather than a vehicle for a foreign gene. Stem cells have an extensive capacity for self-renewal and are multipotent, giving rise to neurons, astrocytes, and oligodendrocytes (Ostenfeld and Svendsen, 2003). The nervous system does not have the regenerative potential of other cell types, making stem cells a potential therapy for diseases and injuries of the nervous system.

Stem cell and gene therapy are powerful research methods showing promise in animal studies of Parkinson's disease (Isacson et al., 2003; Sanchez-Pernaute et al., 2001), lysosomal storage disorders (Jung et al., 2001), stroke (Savitz et al., 2003), retinal degeneration (Chacko et al., 2003), and alcoholism (Thanos et al., 2001).

### Animal Care and Use Concerns Associated with Stem Cell and Gene Therapy

There are unique animal-care issues related to stem cell and gene therapy. Although the brain is relatively isolated from the immune system, immune and inflammatory responses do occur when viral gene therapy is used (Thomas et al., 2001). The most common viruses utilized as gene therapy vectors are lentiviruses (Quinonez and Sutton, 2002), herpes simplex viruses (Burton et al., 2002), adenoviruses (Lai et al., 2002), and adeno-associated viruses (Lai et al., 2002). However, newer generations of viral vectors seem to provoke less serious immune responses (Anonymous, 1996).

Stem cells have also been shown to cause an adverse immune response. This immune response is termed graft-versus-host disease and can occur acutely or chronically in a large percentage of patients (Abo-Zena and Horwitz, 2002). Clinical signs of an immune reaction depend on the species of animal, the type of reaction, and the organs affected. The reactions may result in local or systemic symptoms, including such vague symptoms as fever, vomiting, diarrhea, ataxia, and behavior changes and such dramatic symptoms as anaphylactic shock (Aiello, 1998c) or degeneration of the target organ (Yang et al., 1994). Many times immunosuppressive drugs or irradiation is used in combination with stem cell therapies. These can have significant adverse consequences on an animal's health and well-being, including causing opportunistic infections and cancers due to the immune suppression (Junghanss and Marr, 2002).

Gene therapy also has the potential to be tumorigenic (Donsante et al., 2001) and stem cells have tumorigenic tendencies (Le Belle and Svendsen, 2002; Ruiz et al., 2002). Stem cell transplantation into the brain has also been shown to result in hyperplasia and atypical integration (Zheng et al., 2002). As a result, animals that undergo stem cell or gene therapy should be monitored acutely for immune reactions and chronically for tumor development and neurological dysfunction caused by hyperplasia or atypical integration. A plan for monitoring expected and unexpected consequences should be developed (see Chapter 3).

## Occupational Health and Safety

The potential for unexpected consequences of gene therapy extends to the potential for infection of researchers, animal-care technicians, and other laboratory-animal species with the recombinant DNA under investigation. To manage the potential risks, NIH produced *Guidelines for Research Involving Recombinant DNA Molecules* (NIH, 1998). That document identified which kinds of experiments involving recombinant DNA required institutional biosafety committee approval or notification. Some of the designated experiments include research involving transgenic rodents and the use of infectious DNA or RNA viruses. Appendix Q of the document identifies the physical and biologic containment requirements for handling animals involved in recombinant-DNA research.

# 5

# Prolonged Nonsurvival Studies

Most prolonged nonsurvival studies are carried out in anesthetized animals and extend over many hours or days. In a prolonged nonsurvival study of the visual system, the scientific needs of the experiment usually require that stationary images or images whose motion can be controlled accurately and repeatedly be presented to the retina, so self-generated movements by the animal must be minimized. Minimizing self-generated movements is also required in other neuroscience studies. Neuroscientists frequently resolve this problem by administering neuromuscular blocking drugs (NMBDs) that paralyze all voluntary muscles, including the extraocular muscles (Flecknell, 1987; Hildebrand, 1997). Many of the typical indicators of anesthetic depth (such as response to noxious stimuli and changes in respiratory rate) are thus eliminated, and this makes it difficult to assess whether an animal is experiencing pain and/or distress. But that assessment is critical, both for the welfare of the animal and to avoid compromising experimental results (Moberg, 1999). An earlier workshop on anesthesia and paralysis in experimental animals (Anonymous, 1988) considered the special problems associated with this approach, and they are also discussed in *Preparation and Maintenance of Higher Mammals During Neuroscience Experiments* (NIH, 1991), the predecessor of the present document. Without exception, the scientific need to use NMBDs must be explained in an investigator's animal-use protocol and approved by the IACUC (NRC, 1996).

Principle V of the US Government Principles (IRAC, 1985) states that "surgical or other painful procedures should not be performed on unanesthetized animals paralyzed by chemical agents," and the *Guide* further states (p. 65):

> Neuromuscular blocking drugs . . . do not provide relief from pain. They are used to paralyze skeletal muscles while an animal is fully anesthetized. They

might be used in properly ventilated conscious animals for specific types of nonpainful, well-controlled neurophysiologic studies. However, it is imperative that any such proposed use be carefully evaluated by the IACUC to ensure the well-being of the animal because acute stress is believed to be a consequence of paralysis in a conscious state and it is known that humans, if conscious, can experience distress when paralyzed with these drugs (NIH, 1991; NRC, 1992).

The special concerns associated with prolonged nonsurvival experiments were well summarized in *Preparation and Maintenance of Higher Mammals During Neuroscience Experiments* (NIH, 1991):

> The most critical issues in prolonged nonsurvival experiments arise in the context of anesthesia, maintenance of physiological state, and monitoring of the animal's condition. The choice of anesthetic must jointly satisfy the need of the experimenter to perturb the preparation as little as possible and his/her obligation to ensure that the animal remains free of pain and distress. Maintaining an anesthetized (and often immobilized) animal in sound physiological condition for several days is a considerable challenge and monitoring both the anesthesia and the animal's general condition requires careful attention to a number of kinds of measurement.

A variety of experimental protocols have been used to minimize the difficulties. The *Guide* should be interpreted as a flexible document in reviewing protocols of this sort, because procedures may vary with species and among different experimental paradigms.

The chief animal welfare concern associated with anesthetized paralyzed animals is that the behavioral indicators of pain and/or distress are inhibited by NMBDs, and this makes it necessary to use special measures to monitor and regulate anesthesia (Gibbs et al., 1989). Anesthesia must be regulated in such a manner that it exerts either no effect or a minimal and constant effect on the neurophysiologic responses being measured. Of both animal welfare and scientific concern is the problem of monitoring and maintaining the animal's physiologic state, particularly in experiments that extend over several days (Lipman et al., 1997). Issues can also arise regarding infection, in that it usually is not possible to conduct prolonged nonsurvival neuroscience experiments aseptically and the duration of these experiments is sufficient to allow infections to develop.

A two-step paradigm is used in most prolonged nonsurvival neuroscience experiments. In one kind of study, during an initial step of 2–4 hours duration, the animal is surgically prepared for a subsequent data collection step, which follows immediately and can last for a few hours to several days (NIH, 1991). During the initial surgical preparation step, all procedures are completed under surgical anesthesia without NMBDs, and analgesics may be administered preemptively to augment the anesthetic regimen (see "Anesthesia and Analgesia" in Chapter 3). Other studies employ a variant of the two-step paradigm that involves performing, several days before the nonsurvival recording session, a survival surgery step during which various devices are implanted (such as a cranial pedestal and chamber that are used to secure the animal and the recording device, respectively) (NIH, 1991) (see "Multiple Survival Surgeries" in Chapter 4). The latter approach frequently allows the neuroscientist to perform the initial-survival surgery step with aseptic techniques in a dedicated surgical facility. Furthermore, only minor procedures (such as venipuncture and endotracheal intubation) are needed on the day of the prolonged nonsurvival recording session, and they can be performed with only light anesthesia augmented by analgesics. Following these minor procedures, general anesthesia is provided for the nonsurvival recording session.

The difficulty involved in assessing whether paralyzed animals are free of pain and/or distress is acknowledged by the *Guide*, which states (p. 65):

> Some classes of drugs—such as sedatives, anxiolytics, and neuromuscular blocking agents—are not analgesic or anesthetic and thus do not relieve pain; however, they might be used in combination with appropriate analgesics and anesthetics. Neuromuscular blocking agents (e.g., pancuronium) are sometimes used to paralyze skeletal muscles during surgery in which general anesthetics have been administered (Klein, 1987). When these agents are used during surgery or in any other painful procedure, many signs of anesthetic depth are eliminated because of the paralysis. However, autonomic nervous system changes (e.g., sudden changes in heart rate and blood pressure) can be indicators of pain related to an inadequate depth of anesthesia.

It follows that adequate anesthesia must be established and verified before the administration of NMBDs and initiation of the data-collection session. The animal should be maintained without NMBDs at a fixed anesthetic level until it is physiologically stable. It will take at least 30 minutes, depending on the duration of action of the anesthetics used during the initial surgical preparation period (if there was one), at a fixed anesthetic level and without change in physiological state to ensure that the animal is stable. That period should be used to establish and validate the physiological signs that will be monitored under paralysis to document that the animal is being maintained in a suitable condition. Experience has shown that care should be taken to ensure that the level of anesthesia established during this initial period is adequate but does not compromise neural responsiveness in the areas under study (NIH, 1991). That is critical because reducing the level of anesthesia after NMBDs have been administered is problematic. Obtaining a performance based assessment of the adequacy of the new anesthesia level may require that NMBDs be withdrawn to assess skeletal muscle response; however, this can entail difficulties because a long period may be needed to restore muscle responses (Hildebrand, 1997). Noninvasive assessment of neuromuscular function with examination of evoked responses of skeletal muscle to peripheral motor nerve stimulation can facilitate monitoring both the induction of paralysis and the recovery from NMBDs (Hildebrand, 1997).

Anesthesia and general physiologic state should be monitored for each animal during each procedure (Mason and Brown, 1997; NRC, 1996). Notations should include the time, date (if appropriate), drugs or solutions administered, and the name or initials of the person making the entry. A number of physiologic measures are helpful in monitoring animals on NMBDs, including heart rate, electroencephalogram, arterial blood pressure, blood oxygen saturation, urine production and pH, end-tidal $CO_2$ and/or blood gas concentration, rectal temperature, and general autonomic signs of arousal, such as salivation, pupil size, and lacrimation (Hildebrand, 1997; NIH, 1991). Physiologic measures should be documented periodically throughout an experiment. The periodicity of monitoring and documentation should be more frequent during the initial stages of experimentation (such as every 15 minutes) and gradually extended once the animal has been stabilized on NMBDs. The details of the specific physiologic measures to be monitored and the frequency and means of documentation should be described in the research protocol and approved by the IACUC. The use of automated multifunction measuring devices—which maintain a running record of such metabolic measures as blood pressure, blood oxygenation, expiratory $CO_2$, and pulse rate—can greatly facilitate the monitoring of anesthesia and general physiologic state (Vogler, 1997). However, not all devices maintain a historical record, and regular measurements should be recorded under these circumstances. The use of automated devices cannot substitute for direct monitoring of the animal by a human observer (see also Mason and Brown, 1997), and a human observer should be present at all times during a prolonged nonsurvival procedure, as the clinical status of the animal can change quickly and require intervention. Monitoring data should be filed by experiment and animal and kept for at least the duration of the overall project (NIH, 1986).

The issues involved in maintaining an experimental animal in good physiologic condition during a prolonged recording experiment are similar to those involved in other situations that require the long-term maintenance of animals in clinical situations (NIH, 1991). Animals that are paralyzed must be ventilated artificially, and standard veterinary practices should be followed when selecting the gas mixture and anesthetic or analgesic used (Vogler, 1997). In some cases, gaseous anesthetics or analgesics are included in the gas mixture (such as isoflurane and $N_2O$), while in other cases, room air with or without added oxygen is used. The use of a mixture of 50% $O_2$ and 50% $N_2O$ may be helpful, in that $N_2O$ potentiates intravenous anesthetics (NIH, 1991). However, $N_2O$ increases cerebral blood flow, and this side effect may be of special concern to researchers performing intracranial procedures (Drummond et al., 1987). Animals exposed to artificial ventilation for long periods should be hyperinflated (sighed) at regular intervals to help to avoid collapse of the pulmonary alveoli (atelectasis) (Vogler, 1997). Hydration of the inspired gases is also helpful in preventing desiccation of lung tissues. That can be accomplished by passing the air through fluid in a flow-through system or by using a closed-loop system that obviates additional hydration; passive humidification devices are also effective, inexpensive, and appropriate for multi-day use. A straightforward way of monitoring the adequacy of artificial ventilation is to measure end-tidal $CO_2$ either continuously or at frequent intervals (Vogler, 1997).

Animals are typically given an osmotically balanced fluid and metabolites, such as lactated Ringers solution with dextrose, intravenously (DiBartola, 2000; Haskins and Eisele, 1997). When data collection goes on for several days, supplementation of that solution with potassium and/or amino acids may be desirable. For shorter experiments (under 48 hours), periodic subcutaneous administration of fluid may be sufficient. In all cases, the total volume administered should be adequate to make up for what is lost through the skin and lungs and should be sufficient to maintain renal function. The role of the kidneys in maintaining pH and osmotic balance is critical, and normal renal function can be particularly important in preventing the imbalances that may occur when animals are subjected to prolonged artificial ventilation. Monitoring urine output may be helpful in some situations to ensuring adequate renal function and hydration.

During long experiments, rearranging an animal's limbs and body and massaging the large muscle masses regularly can help to prevent the edema and venous pooling that occur in the absence of muscle tone and movement. Providing regular doses of antibiotics, vitamins, and anti-inflammatory agents may help to keep an animal in a stable condition and prevent infection (NIH, 1991). Core body temperature should be monitored throughout the period of paralysis, and supplemental heat should be provided as needed with forced air or circulating water heating pads (e.g., Vogler, 1997).

Evaluating the need for aseptic technique in prolonged nonsurvival experiments requires the professional judgment of the investigator, veterinarian, and IACUC case by case. In general, the need for asepsis will depend on the duration of the experiment and the extent to which it involves the exposure of tissues or body cavities. As stated in APHIS/AC Policy 3, "nonsurvival surgeries require neither aseptic techniques nor dedicated facilities if the subjects are not anesthetized long enough to show evidence of infection." Any procedure that lasts longer than 12 hours and involves exposed tissues or body cavities presents a significant opportunity for infection to occur, and the risk increases with the length of the procedure (Knecht et al., 1987; McCurnin and Jones, 1985).

Failure to use aseptic procedures in prolonged nonsurvival experiments increases the possibility that research data will be compromised and increases the risk of premature death due to sepsis. Either of those outcomes could entail the use of additional animals, which would be inappropriate for both ethical and regulatory reasons (see U.S. Government Principle III). It would also violate the requirements of the *Guide* and the AWRs for the provision of adequate veterinary care. The use of aseptic procedures also helps to ensure that students learn proper surgical technique and strengthens an institution's ability to respond to public inquiries about the use of animals in research.

When a preparatory survival surgical procedure is conducted to implant devices, such as a pedestal and chamber, it should be performed under aseptic conditions (AWR 2.31(d)(1)(ix); NRC, 1996). Most implanted devices and their carriers can be disinfected, but it might not be possible for some sensitive or delicate equipment, such as some types of microelectrodes (for further information on this topic see "Animal Care and Use Concerns Associated with Introduction of Probes into Neural Tissue" in Chapter 4). Whenever possible, it is advisable to sterilize or disinfect devices before their insertion into neural tissue. Because a typical neuroscience laboratory contains many other items, such as recording equipment, that cannot be sterilized, the full application of aseptic technique during a prolonged nonsurvival experiment is usually impossible. One approach to that problem is to use appropriate aseptic technique to create and maintain a local sterile field that includes any openings into major body cavities that are made during a prolonged nonsurvival session.

Institutions should develop policies and guidelines to assist investigators in adapting aseptic surgical procedures to the laboratory setting. Topics that should be considered in preparing guidelines include preparation of the laboratory room, with particular attention to the site where surgery and recording will take place (for example, taking into account the relative locations of supply and exhaust ventilation ducts with respect to airborne contamination of the surgical field); preparation of the animal; preparation of the surgeon and any other experimenters who will come into proximity to the animal; instrument preparation; intraoperative monitoring; and training (APHIS/AC Policy 3; NRC, 1996, pp. 78–79). Neuroscientists can assist veterinarians and IACUCs in developing performance-based standards for monitoring the occurrence of deleterious effects by providing postmortem tissue specimens for histopathologic analysis.

# 6

# Studies of Neural Injury and Disease

## DISEASE MODELS

A wide variety of animal disease models are used in neuroscience research to study the causes and treatments of neurologic and psychiatric diseases. They include models of degenerative diseases (such as Alzheimer's disease, Parkinson's disease, and frontotemporal dementia); of traumatic injury of the head, spinal cord, or peripheral nervous system; of infectious diseases (such as immunodeficiency viruses, prion diseases, and viral, bacterial, or parasitic meningitis or encephalitis); of neuroimmunologic disorders (such as multiple sclerosis, myasthenia gravis, and polymyositis); of neurodevelopmental disorders (such as autism, Asperger's syndrome, and Williams' syndrome); of pain (from tissue injury and nerve injury); of neurologic problems that are secondary to primary medical conditions (such as diabetic neuropathy, nutritional disorders, and hepatic and renal encephalopathy); and of psychiatric disorders (such as schizophrenia and affective disorders).

Major considerations in the evaluation of research protocols and the management of animals experiencing those conditions are assessment of animal well-being, provision of appropriate nursing care and pain management, limitation of the duration and severity of the condition to be consistent with the experimental goal, and, in some cases, assessment and minimization of potential human health risks.

Assessment of animal well-being is discussed in detail in Chapter 2, and the same principles apply here. Personnel who are knowledgeable about the species-typical behavior of the animals under study and the clinical symptoms of the disease should evaluate animal well-being at appropriate intervals. Clinical symptoms could involve such diverse markers as musculoskeletal abnormalities (tremor, reduced ambulation, and paralysis), decreased appetite (anorexia or aphagia), adipsia, signs of pain, fever, seizures, disorientation, and/or self-mutilation.

Before initiation of the research project, the research team, in consultation with the veterinary staff, should determine a course of clinical intervention or management based on the observed or expected clinical signs (see Table 6-1). The clinical plan should prevent the development of unintended pain and distress; but in instances where pain and/or distress is an intended outcome (as in pain research), the adverse consequences to the animal should be minimized as much as possible without jeopardizing the research goals. Potential interventions include the provision of easy access to water and perhaps to highly palatable food, promotion of urination and defecation, avoidance of decubital ulceration, maintenance of fluid balance, administration of appropriate analgesic or tranquilizing drugs, and, for some species, human contact to soothe and comfort the animal. Close clinical observation may also be necessary during periods of disease exacerbation. For example, in studies of seizure induction or treatment, continuous animal observation during the seizure is essential to prevent injury to the animal, although the seizure itself may not be painful.

The duration and severity of the condition should be managed within predetermined limits (humane endpoints) that reflect experimental goals. For example, studies of the mechanisms of disease development might require a shorter post-procedural duration than would studies of treatment interventions. Endpoints should be defined in terms of both the experimental goals (such as development of the syndrome or recovery of function according to some objective or subjective standard) and animal well-being (such as clinical deterioration that indicates that euthanasia is warranted). Endpoints applied to a specific model can be modified as more is learned about the model. For example, although death was used as an endpoint in some early studies of tumor metastasis in mice, later studies used hind limb paralysis to indicate that euthanasia should be performed, because paralysis was shown to be a valid indicator of death (Huang et al., 1993, 1995). Seizure studies should be designed to minimize the number, duration, and severity of seizures, without jeopardizing the scientific goals of the study.

## Occupational Health and Safety

The production of some animal disease models requires the use of substances that pose health risks to humans. They include neurotoxins (such as MPTP), infectious agents (such as prions, bacterial, and viral pathogens), human cell lines, and noise exposure (free field). Appropriate risk assessment and the development of safe standard operating procedures is essential for the review and use of these models.

TABLE 6-1 Animal Welfare Considerations Associated with Disease Models[a]

| Type of Disease | Potential Problems | Interventions | Occupational Health Issues |
|---|---|---|---|
| Degenerative (Parkinson's disease, Alzheimer's disease, frontotemporal dementia) | Adipsia, aphagia, tremor, incoordination, disorientation | Nursing care | MPTP |
| Trauma (head and spinal cord injury, peripheral nerve repair and regeneration) | Paralysis, distress, self-mutilation, seizures | Nursing care, local anesthetics, analgesia, sedatives, antibiotics, anti-convulsants | |
| Infectious disease (HIV, prions, meningitis, encephalitis) | Seizures, debility, paralysis, disorientation | Monitoring, antibiotics, anti-convulsants | Microbial pathogens |
| Brain and spinal cord tumors | Pain, paralysis, disorientation, seizures | Analgesia, nursing care, anti-convulsants | Human cell lines |
| Stroke | Surgery complications, paralysis, disorientation, visual deficits | Monitoring, nursing care | |
| Seizures (genetic, audiogenic, drug- or lesion-induced) | Kindling, status epilepticus, secondary injury | Monitoring, anti-convulsants | Noise exposure (free field) |
| Immune-mediated (multiple sclerosis, myasthenia gravis, polymyositis) | Muscle weakness, paralysis | Nursing care | |
| Secondary medical conditions (diabetic neuropathy, renal or hepatic encephalopathy) | Self-mutilation, disorientation | Nursing care | |
| Pain (peripheral nerve damage, tissue damage, central nervous system damage) | Self-mutilation, guarding, changes in locomotor activity, aggression | Local anesthetics, analgesics, sedatives, environmental management | |
| Psychiatric diseases (depression, schizophrenia) | Learned helplessness, social withdrawal | Monitoring, environmental enrichment | |

[a] These lists are intended to be representative, not all-inclusive.

## LESIONS

Neuroscientists often induce lesions to learn about normal brain function (NIH, 1991). A classic example involves the study of the hippocampus and recognition memory, in which rats and monkeys with lesions limited to the hippocampus are impaired in tests of recognition memory (Zola and Squire, 2001).

Lesions of the nervous system can be produced by various means. Surgical or vacuum ablations, stereotaxic administration of neurotoxins, electric lesioning, and vascular occlusions require opening the cranial cavity and are considered major survival surgery that requires aseptic technique and, depending on the circumstances, the use of dedicated facilities (NRC, 1996). Noninvasive techniques can include radiation, blunt trauma, and intravenous administration of neurotoxins, although some of these methods may also be applied directly to the brain after surgical opening of the skull.

The use of lesions can establish an essential role of a structure, but because the processes of learning and memory have many steps, permanent lesions cannot be used to determine which step of learning and memory depends on the structure under study. To address the latter question, reversible lesions are used: the target structure can be temporarily deactivated during different stages of the assessment of learning and memory (for example, at the time of initial learning, during the delay interval, or at the time of retrieval).

Reversible lesions now allow assessment of cognitive function during different phases of learning and memory. Two types of reversible human amnesias have been studied in animal models using reversible lesions: transient global amnesia (Kritchevsky and Squire, 1989) and the amnesia associated with electroconvulsive therapy (Squire, 1986). Three approaches for making reversible lesions are cooling, chemical treatments, and transcranial magnetic stimulation (Lomber, 1999). The first two involve placing implants in the brain (cooling probes or cannulae). The use of implants to produce reversible lesions should be consistent with the guidelines of asepsis and sterility previously discussed ("Animal Care and Use Concerns Associated with Introduction of Probes into Neural Tissue" in Chapter 4).

In some studies, lesions are induced to produce animal models of naturally occurring diseases. For example, a lesion of the nigrostriatal pathway leads to motor deficits associated with Parkinson's disease and was helpful in development of new dopaminergic agents for treatment and in demonstrating the effectiveness of neural transplantation (Tolwani et al., 1999). Such a lesion is usually produced by stereotaxic injection of 6-hydroxydopamine or intravenous administration of MPTP. Parkinsonian lesions have been induced in various animals, including mice, rats, cats, dogs, sheep, and nonhuman primates (Zigmond and Stricker, 1989). The clinical symptoms of the lesions depend on the species used but can include hypokinesis, circling behavior, aphagia, adipsia, bradykinesia, rigidity, balance impairment, and resting tremor (Schneider and Kovelowski, 1990; Stern and Langston, 1985; Taylor et al., 1997).

A common strategy in neuroscience research is to induce lesions that produce specific structural or functional deficits. The deficits then can be studied to develop treatments that lead to recovery of function, as in spinal cord or peripheral nerve injury research. Common, clinically relevant lesion models include spinal cord contusion (Allen, 1911; Dohrmann et al., 1978; Wrathall et al., 1985), spinal cord transection (Khan et al., 1999), and neurectomy (Bouyer et al., 2001). Neurectomy and deafferentation surgery can result in autotomy (self-mutilation of the denervated limb) (Blumenkopf and Lipman, 1991); however, the use of local anesthetic at the time of nerve section will reduce autotomy (Magnusson and Vaccarino, 1996), and may be appropriate if the study of autotomy or dysesthesia is not the goal of the experiment.

## Monitoring and Care Plan

Each nervous system lesion model has the potential for unique animal care issues that need to be fully investigated so that a monitoring and care plan can be drawn up before experimentation.

Preoperative health assessment and postoperative care are particularly important in lesion studies. In many cases, it is also useful to document the performance of an animal in a behavioral task to create a baseline assessment. For example, changes in exploratory behavior can be used as measures of chronic pain after spinal cord contusion in rats (Mills et al., 2001), although such measures alone are not sufficient to define an experimental manipulation as nocifensive.

The perioperative care and monitoring of an animal with a lesion are similar to standard surgical care and monitoring. As suggested by the *Guide*, surgical monitoring may include monitoring core temperature, cardiovascular and respiratory function, and postoperative pain or discomfort (p. 63). A heightened level of monitoring is beneficial in these models to determine whether the lesion caused unusual or unexpected pain and/or distress postoperatively. The difficulty of predicting whether a lesion will compromise an animal's health or well-being (NIH, 1991) reinforces the need for frequent and comprehensive monitoring.

## Impaired Physiologic Functioning

A lesion may cause changes in physiologic function. For example, animals with thoracic spinal cord transection often need to have their bladders manually expressed, and can have inhibited gastrointestinal motility. Close monitoring of bowel and urinary bladder status of these animals is required; additional care, such as the administration of laxatives (Khan et al., 1999), may be necessary in some cases. Information about the physiologic difficulties associated with well-characterized models is readily available in the literature and should be incorporated into the long-term care plan described in the animal-use protocol. In new models that are being developed, a specific monitoring plan should be developed to assess possible changes in physiologic status.

### Reduced Capabilities

Lesions that impair animals' mobility or alter their motivation can reduce their ability to care for themselves. They may be anorexic or adipsic and may not groom adequately. Adequate monitoring (scheduled weighing and assessment of appearance) and detailed record keeping will alert researchers and animal care staff to administer extra care (NIH, 1991), possibly even euthanasia in accordance with predetermined endpoint criteria. To ensure the consistency of monitoring between observers and experiments, it may be beneficial to use a quantitative scoring system for monitoring appearance and grooming (Ullman-Cullere and Foltz, 1999).

The additional animal care provided may include administration of nutritional support (Ungerstedt, 1968), fluid replacement, or provision of soft, rather than hard, food (NIH, 1991). In some instances, such as with stereotaxic injections of 6-hydroxydopamine, the lesion may be administered unilaterally rather than bilaterally. That approach is sometimes adequate for research purposes and reduces the impairment (Tolwani et al., 1999).

## ANIMAL MODELS INVOLVING PAIN

The approaches used to recognize and treat unintended pain originating in neuroscience studies and in studies of pain are basically the same and are discussed elsewhere ("Pain and Distress" in Chapter 2). This section focuses on models of inflammation and nerve injury that produce pain so that its underlying mechanisms can be studied. Experiments with animals have mostly used stimuli that produce acute pain of short duration and moderate intensity; these models have become standards in the screening of putative analgesics. More recently, investigators have begun to develop nonhuman animal models that mimic persistent pain conditions seen in humans. Tissue injury and inflammation are commonly associated with clinical conditions that lead to persistent pain. Accordingly, new animal models to study these conditions differ in important ways from earlier, acute pain models.

Animal models of pain and hyperalgesia (excessive sensitivity to pain) have been developed to study the functional changes produced by the injection of inflammatory agents into the rat or mouse hindpaw (for review see Ren and Dubner, 1993; Dubner and Ren, 1999). The animals withdraw their limbs reflexively but also exhibit more complex organized behaviors, such as paw-licking and guarding (Hargreaves et al., 1988). A paw-withdrawal latency measure and withdrawal duration can be used to infer pain and hyperalgesia in response to thermal or mechanical stimuli (Ren and Dubner, 1993). Methods of measuring nocifensive behavior have also been applied to the orofacial region (Imamura et al., 1997; Ren and Dubner, 1993). In the above studies, most of the nocifensive behaviors provide an animal with control of the intensity or duration of the stimulus in that the behaviors result in removal of the aversive stimulus.

Animals in persistent-pain models do not have control of stimulus intensity or duration. For example, the writhing response is produced in rodents by injecting pain-producing chemical substances intraperitoneally. The acute peritonitis resulting from the injection produces a response characterized by internal rotation of one foot, arching of the back, rolling on one side, and accompanying abdominal contractions. The writhing response is considered a model of visceral pain (Vyklicky, 1979). Not only does the animal lack stimulus control with this method, but the experimenter cannot control the duration of the stimulus. In another test, formalin is injected beneath the footpad of a rat or cat (Abbott et al., 1995; Dubuisson and Dennis, 1977). The chemical produces complex response patterns that last for about an hour. Many response measures are used for assessing pain after formalin injection. They include single measures such as flinching, shaking, and jerking—or complex scores that are derived from several nocifensive behaviors, such as licking or guarding (Clavelou et al., 1995). However, the animals do not have complete control over the aversiveness of the persistent stimulus. Vocalization is another common, unlearned reaction to painful stimuli (Kayser and Guilbaud, 1987), and the stimulus intensity necessary to elicit a vocal response from the animal can be determined. The stimulus can be applied to any part of the body; again, the animals cannot control the intensity or duration of the stimulus.

Nerve-injury models that mimic neuropathic pain in humans have been developed recently (Dubner and Ren, 1999). Partial nerve injury in the rat results in signs of hyperalgesia and spontaneous pain. In one model, loose ligatures are placed around the sciatic nerve; demyelination of the large fibers and destruction of some unmyelinated axons result (Bennett and Xie, 1988). In another model, ligation and severing of the dorsal one-third to one-half of the sciatic nerve produce similar behavioral changes (Seltzer et al., 1990). Kim and Chung (1992) have developed a third model, in which the L5 and L6 spinal nerves are tightly ligated on one of the rat's sides. All three models mimic clinical conditions of painful neuropathy and yield evidence of persistent spontaneous pain, allodynia (pain resulting from a nonnoxious stimuli), and hyperalgesia. These nerve-injury models of neuropathic pain have been adapted for use in mice (Malmberg et al., 1997; Ramer et al., 1997), in which they can be used to study pain mechanisms in transgenic models.

### Ethical Considerations Associated with Pain Research

Anesthetic and pain-relieving methods and drugs generally act on the system under study—the nervous system—and neuroscientists and IACUCs must make difficult choices in selecting the means by which pain and distress are controlled and how much pain and/or distress is acceptable.

Several ethical issues have been proposed for IACUC consideration when reviewing protocols involving pain and/or distress in animals (Tannenbaum, 1999):

- Of the animal-use protocols reviewed by the IACUC, those which include pain and/or distress should be subject to a full committee review rather than review by a designated member or with an expedited review process. If necessary, the committee should involve an outside consultant to understand better the ramifications of the study.
- The protocol should provide a compelling justification for the work, a description of the qualifications of the personnel who will perform the work and provide care for the animals, and a rationale for withholding analgesics or other pain-relieving or distress-relieving methods.
- The protocol should contain a complete and accurate description of the severity of pain and/or distress that will potentially be experienced by the animals.
- When it is not in conflict with the scientific goals of a well-designed study, pain relief should be provided by anesthetizing the animals; giving them analgesics; allowing them to escape or avoid the pain; or control the experimental trials.
- A humane endpoint for the use of an animal should be determined as an element of the protocol, before work begins.

Ethically, models of persistent pain present a particular challenge because they produce pain that most guidelines for the use of animals in research state should be avoided. Scientists should demonstrate a continuing responsibility for the proper treatment of the animals involved in these experiments. Because some models produce persistent pain that the animals cannot control, it is important that investigators assess the level of pain in these animals and provide analgesic agents when they do not interfere with the purpose of the experiment. A reduction in body weight or a significant deviation from normal behavior—such as a change in normal activity patterns, social adjustment, feeding behavior, and sleep-wake patterns—suggests that an animal is in severe and possibly intolerable pain.

In animal models of inflammation and nerve injury, the IACUC should ensure that steps are taken to safeguard animal welfare. The steps may include use of fail-safe devices to avoid excessive exposure to painful stimuli (for example, monitoring stimulus intensity and duration), having in place well-established humane endpoints to deal appropriately with intractable conditions (such as self-mutilation), and postprocedure monitoring of animal well-being.

# 7

# Perinatal Studies

Research programs involving perinatal (fetal and neonatal) animals offer insight into the development of the brain and central nervous system and the age-dependent effects of genes, toxicants, and the environment. Such research has shed light on autism, learning disabilities, and fetal alcohol syndrome. Experiments involving perinatal animals use the same techniques discussed in other chapters and therefore pose the same animal care and use concerns as experiments involving adult animals. However, three issues influence how the care and use concerns are addressed in perinatal studies (NIH, 1991). First, perinatal physiology can be radically different from adult physiology, and it changes throughout early development, affecting such aspects of animal use as appropriate euthanasia and analgesia. Second, perinatal studies often entail the use of animals at a variety of developmental stages, which differ physiologically, as in the permeability of the blood-brain barrier (Saunders et al., 2000); this may necessitate that animals of different ages be cared for differently even if they are used in the same experiment. Third, when fetal studies are proposed, the welfare of both the mother and the fetus must be considered.

## DEVELOPMENT OF PAIN PERCEPTION

The development of neural systems essential for pain perception has been studied most extensively in the rat, as has the early development of pain-related behaviors. Although there is no definitive evidence that prenatal animals can *perceive* pain, reflexive behavior in fetal animals sometimes correlates with behavior exhibited by adult animals in response to pain stimuli. It is not known when developing animals begin to perceive pain. Reflexive withdrawal from noxious stimulation is observed in rodent embryos starting in late gestation, for example, on embryonic day 17 (E17) in the rat fetus (Narayanan et al., 1971). Human fetuses develop stress hormonal and circulatory changes in response to noxious stimuli by 18–20 weeks of gestation; similarly, fetal lambs and rhesus monkeys demonstrate changes in the pituitary-adrenal axis after application of stressors at the late gestation ages of 125 days and 133 days, respectively (Rose et al., 1978; Smith et al., 2000). Behavioral responses to injection of an irritating substance (formalin) into the paw can be seen in rat fetuses as early as E19, and the response correlates with expression of the *c-fos* protein (an indication of neuronal activation) in the spinal cord by E20 (Yi and Barr, 1997). By birth, neural substrates for perception of noxious stimulation are present in the periphery and spinal cord of the rat pup, although sensory systems are immature and undergo substantial change during the first few weeks after birth. Many neurotransmitters and receptors in pain pathways appear early in development, but their expression may vary—in either direction—during the neonatal period and may take weeks to achieve adult levels. One could argue that the physiologic response to noxious stimuli suggests a correlation with sensitivity to pain (Mahieu-Caputo et al., 2000; Smith et al., 2000). In theory, cortical recognition of pain in a human fetus should occur in the 26th week of gestation with development of thalamocortical connections (Vanhatalo and van Nieuwenhuizen, 2000).

Rat pups show behavioral arousal and withdrawal responses to noxious thermal and mechanical stimuli as early as the first postnatal day (Barr et al., 1992; Blass et al., 1993; Fanselow and Cramer, 1988; Fujinaga et al., 2000). In addition to behavioral responses to injection of irritating chemicals (Abbott and Guy, 1995; McLaughlin et al., 1990), rat pups as young as 3 days show allodynia and hyperalgesia in response to experimentally induced inflammation (Marsh et al., 1999). The behavioral responses of young pups to noxious stimulation are mostly generalized, whole-body responses, such as wriggling, although more localized withdrawal responses are also seen. As pups mature, their responses become more organized and localized and more typical of adult responses.

Dampening of behavioral responses to noxious stimuli, particularly when such opioid drugs as morphine are administered, is also seen within the first few days after birth (McLaughlin et al., 1990; Fanselow and Cramer, 1988) and the sedating effects of such drugs as pentobarbital can be distinguished from the analgesic effect of morphine as early as postnatal day 1 (P1) in the rat pup (Abbott and Guy, 1995). Mature responses to analgesics are seen around the age of 3 weeks in the rat, coinciding with the maturation of supraspinal descending inhibitory processes. Little information is available regarding neonatal precocious mammal responses to analgesics during postnatal development; however, most neonatal animals develop physiologic responses that are consistent with adult responses by the age of 6–8 weeks. In addition, many physiologic differences between neonatal and adult animals—such as the neonate's greater permeability of the blood-brain barrier, higher body water content, less mature hepatic microsomal enzyme systems, and lower albumin concentrations—affect the pharmacodynamics and pharmacokinetics of analgesic and anesthetic drugs (Thurmon et al., 1996).

In summary, although there is not enough evidence to determine whether neonatal animals perceive pain, some stimuli that are noxious to adult animals have been shown to trigger reflexive behavior in neonates, and this suggests that neonates would benefit from the administration of analgesics.

## ANESTHESIA AND ANALGESIA

### Fetus

Available evidence suggests that the late-term fetus (E19–E20 in the rat) is responsive to noxious stimulation, as is the late-term fetal lamb and 26-week human fetus. Therefore, provision of anesthesia for potentially painful procedures is advised for late-term fetuses. For fetal manipulations in utero, anesthetics used to prevent pain in the mother are probably adequate to prevent pain in the fetus. Most drugs used for anesthesia in mammals—including barbiturates, ketamine, opioids, and inhalant anesthetics—readily cross the placenta. Therefore, the primary consideration should be adequate anesthesia, analgesia, and supportive care for the dam.

Anesthetic agents widely accepted for use in fetal surgical procedures include such inhalants as halothane, isoflurane, and desflurane (Abboud et al., 1995; Sabik et al., 1993). Balanced anesthesia with isoflurane and thiobarbiturates has been successfully used for late-term fetal pigs (Sims et al., 1997), whereas methoxyflurane and xylazine are associated with postnatal mortality in puppies delivered by caesarean section when those drugs were used for anesthesia in the dam (Moon et al., 2000).

Monitoring of anesthesia in fetal animals presents several challenges. Electrocardiographic monitoring can be used to easily monitor heart rhythm and electrical activity in fetuses of larger mammals, though bradycardia is a poor indicator of fetal distress. Pulse oximetry is noninvasive and effective for fetal lambs. It has a rapid response and is simple to use on the exposed fetus of larger mammals (Luks et al., 1998b). Direct monitoring of blood pressure and intravascular oximetry can quickly and accurately indicate fetal distress but are generally considered impractical because of their invasiveness.

In some studies, consideration of the potential effect of in utero drug exposure on physiologic and behavioral development of the animal may be appropriate (e.g., Belcheva et al., 1994; Niesink et al., 1999; Rodier et al., 1986). Nonopioid analgesics such as acetylsalicylic acid and acetaminophen, are potent inhibitors of prostaglandin synthesis, and their use in a fetus may result in unintended physiologic effects (Peterson, 1985). Prenatal administration of meperidine or bupivacaine to primates may influence behavioral maturation (Golub, 1996). In higher mammals, such as nonhuman primates, appropriate postoperative analgesia for the dam is an important precaution in preventing premature labor after intrauterine surgery (Tame et al., 1999).

For experimental protocols that require the manipulation of late-term rodent fetuses after their removal from nonanesthetized mothers (such as a dam euthanized by decapitation or cervical dislocation), guidelines for anesthesia and analgesia in neonates should be followed.

## Neonate

Potentially painful experimental manipulations in neonatal rodents require the use of anesthesia or analgesia unless the IACUC has approved withholding anesthesia or analgesia for scientific reasons. The primary difficulty in using anesthesia or analgesia in the neonate is balancing its effectiveness and safety. Many anesthetics that can be used safely and effectively in adult rodents are not good choices for neonates; two examples are pentobarbital and ketamine, both of which tend to be ineffective at lower doses and fatal at higher doses (Danneman and Mandrell, 1997). In general, neonatal rodents are more sensitive to anesthetic and analgesic drugs than are adult animals, and such toxic effects as respiratory and cardiac depression are more serious problems in the youngest animals (e.g., Colman and Miller, 2001; Fortier et al., 2001; Greer et al., 1995; Prakash et al., 2002).

Most of the anesthetic agents used in juvenile and adult animals are safe and effective in larger neonatal mammals (Grandy and Dunlop, 1991; Thurmon et al., 1996). The choice of anesthetic agent used may depend on species, type and duration of procedure, and availability of specialized equipment needed (such as a gas anesthesia machine with a precision vaporizer). Most anesthetic regimens used in precocious and nonrodent neonatal mammals are standard veterinary procedures.

The choice of anesthetic method or agent should be based on the procedure, expertise of the researcher, the potential for hemorrhage, and the stability of the anesthetic plane. Overall, the best results of anesthesia in neonatal rodents have been achieved with inhalant anesthetics and hypothermia. Inhalants are a reasonable first choice for anesthesia of neonatal rodents. When inhalants cannot be used—for safety or practical reasons—hypothermia should be considered as a safe and effective alternative to injectable drugs. Hypothermia has been proved safe and effective as the sole method of anesthesia for altricial rodents (such as rats and mice) up to about the age of 7 days (Danneman and Mandrell, 1997; Phifer and Terry, 1986). However, it has the potential to be noxious, and rapid cooling of nonprotected flesh is painful (Wolf and Hardy, 1941). Rat pups recovering from hypothermia—but not pups recovering from general anesthesia—emit ultrasonic vocalizations even when placed with their mothers during the recovery period (Hofer and Shair, 1992). The significance of the vocalizations is not clear, but they may indicate distress. To reduce possible unintended pain associated with cooling, the technique for inducing hypothermia should include partial insulation of the pup (for example, by wrapping in a latex blanket) (Danneman and Mandrell, 1997).

As with adult animals, assessing the effectiveness of anesthesia in neonates is important before beginning a potentially painful procedure. Adequately anesthetized rat pups will not respond to a light pinch of the foot or tail. Similarly, adequately anesthetized adult rats will not respond to a pinch of the toe or tail.

Opioid drugs provide effective analgesia against thermal, inflammatory, and mechanical pain in neonatal rodents as young as P1 (Barr, 1999; Barr et al., 1992; Helmstetter et al., 1988; Marsh et al., 1999; McLaughlin and Dewey, 1994) and should be considered for use whenever analgesia would be provided for an adult animal. Fentanyl is a recommended analgesic for neonatal dogs and humans because it has less of a respiratory depressant effect than morphine (Luks et al., 1998a).

Neonatal exposure to pain, especially when pain is an unintended outcome, may have developmental effects on the central and peripheral nervous systems and alter behavior and the threshold for pain in adulthood (Anand et al., 1999; Bhutta et al., 2001; Fitzgerald and Beggs, 2001).

## SURGERY, POSTOPERATIVE MONITORING, CANNIBALISM, AND NEGLECT

Aside from the technical difficulties associated with using very small animals, surgical procedures involving neonatal rodents present such challenges as maternal neglect and cannibalism. As with adult rodents, pups should be kept warm, dry, and well hydrated postoperatively. They should be placed in a warm—not hot—environment until they have regained the ability to right themselves when placed on their backs or sides, after which they should be returned to their mothers. Some rodent mothers (particularly in some strains such as BALB/c mice) will reject or kill their pups under these circumstances. Some steps can be taken to reduce that problem. First, pups should be sufficiently recovered from anesthesia that they are able to right themselves and respond to stimulation. Smearing a pup with bedding and urine from littermates that remained with the mother can be helpful, as can placing the pup in the middle of the litter and allowing it to settle in for a minute or two before reintroducing the mother. Other methods that may work include masking olfactory cues by sprinkling baby powder on mother, pups, and bedding and smearing the pups and the mother's nose with an aromatic agent, such as Vicks Vapo Rub®.

The following method is cumbersome, but it can greatly improve the rate of successful reunion of mouse pups with their mothers and might be considered when maternal neglect of pups is substantially inhibiting progress of a study:

When pups are removed for surgery, similarly aged pups can be taken from an outbred mouse (such as CD-1) and transferred to the mother whose pups were taken for surgery. Postoperatively, the surgically altered pups are then placed with the outbred mother for temporary fostering during the recovery period and left with her for a couple of hours or overnight. The litters are then switched so that each mother has her own pups back. Care must be taken to treat both experimental and control pups in the same way to avoid introducing experimental variability.

In any event, the mother's behavior toward the pup should be observed closely for the first 10–15 minutes after the pup is returned to her and then every 10–15 minutes for the next couple of hours. At the first sign of aggression by the mother toward the pup, the pup should be removed. If other means of caring for the pup (such as fostering or hand rearing) are not available, the pup should be euthanized, as should pups that are not being cared for by their mother.

In higher mammals, neglect and cannibalism are uncommon postsurgical problems. However, the behavior of the dam should be closely monitored after return of the neonate to her.

## IDENTIFICATION, TAGGING, TATTOOING, AND TOE CLIPPING

Two of the most common methods of identifying adult rodents, ear notching and ear tagging, are not useful for neonatal rodents, because they have small ears tightly placed against their heads. Temporary identification of hairless neonates can be achieved with nontoxic indelible markers (for example, Sharpie®). However, this marking rarely lasts for more than a day, because the mothers will lick the color off. More permanent identification can be achieved by marking the tail with a tattoo machine designed for this purpose; with practice, pups can be marked quickly and effectively. According to the *Guide*, "toe clipping [removal of the first bone of certain toes, corresponding to a predetermined numbering code], as a method of identification of small rodents, should be used only when no other individual identification method is feasible and should be performed only on altricial neonates" (p 46). Under some circumstances, that method of identification may be necessary, but it should be used only with IACUC approval based on appropriate justification in the animal-use protocol.

## REGULATORY CONSIDERATIONS IN FETAL SURGERY

Many experimental fetal surgical procedures in higher mammals require special procedures or conditions, such as a second surgery for the injection of tracers or producing a lesion, or specialized equipment and facilities. Exposure of a fetus in utero constitutes a major operative procedure as defined by the AWRs and the *Guide*. In accordance with regulatory requirements for surgery, multiple survival surgical procedures must be justified scientifically by the neuroscientist in the animal-use protocol and approved by the IACUC. In addition, traditional tracer injections, lesions, or recording may require that the surgical procedure be conducted outside facilities dedicated for aseptic surgery (such as in a laboratory setting). This represents a deviation from the *Guide* and the AWRs, so approval for such procedures rests with the IACUC. Performance standards and a team approach by the IACUC, the veterinarian, and the investigator can ensure that the spirit of the regulation is met and that veterinary care will not be compromised as a result of surgical procedures conducted under non-aseptic conditions (see "Asepsis and Physical Environment" in Chapter 3 and "Modified Surgical Settings" in Chapter 4).

## EUTHANASIA

Laboratory animals can be euthanized in three ways: hypoxia, depression of neural activity necessary for life function, and physical disruption of brain activity and destruction of neurons necessary for life (Balaban and Hampshire, 2001). However, the physiology of the perinatal animal renders some of the euthanasia methods used for adult animals inadequate and therefore inadvisable for perinatal animals (NRC, 1996).

In rodent fetuses that are less than E14, the lack of neural development prevents signs of fetal response to noxious stimuli, so euthanasia of the dam or removal of the fetus from the dam will result in the painless death of the fetus without a requirement for additional measures (NIH, 1997).

Inhalant agents, including inhalant anesthetics and $CO_2$, that cause death by cerebral depression and/or hypoxia, must be used carefully for euthanasia of older fetuses or neonates. The comparatively hypoxic intrauterine environment renders these young animals much more tolerant of hypoxic conditions than adults (Singer, 1999), and euthanasia with an agent that causes death by hypoxia, such as $CO_2$, may take 30 minutes or longer. Therefore, if these agents are used, personnel should be appropriately trained to use prolonged exposure times. Ideally, death should be verified by a secondary method such as decapitation or cervical dislocation.

Older fetuses and neonates can also be euthanized with chemical anesthetics, decapitation, or cervical dislocation. If chemical fixation of the whole fetus is necessary, the fetus should be properly anesthetized before fixation (NIH, 1997). In accordance with the report of the AVMA Panel on Euthanasia (2001), some physical methods of euthanasia, such as decapitation, require appropriate training, experience, and specific approval by the IACUC.

# 8

# Agents and Treatments

This chapter deals with a wide assortment of experimental treatments: drugs and toxicants; exposure to heat, light, or sound; modification of nutrients; induced exercise; and sleep deprivation. Although seemingly dissimilar, those treatments all have the potential for inadvertent injury to experimental animals if the animals are not carefully monitored, especially if stimuli are introduced with a mechanical device, such as a treadmill or hotplate. The first section updates material in the NIH report *Methods and Welfare Considerations in Behavioral Research with Animals* (NIH, 2002, pp. 57–66).

## PHARMACOLOGICAL AND TOXICOLOGICAL AGENTS

Drugs and toxicants are administered for various purposes (Goldberg and Stolerman, 1986; van Haaren, 1993; Weiss and O'Donoghue, 1994). A drug or toxicant may be administered to:

- observe neurobehavioral effects to determine whether a drug can alleviate health problems (such as pharmacotherapy for behavioral and neurologic disorders),
- determine how a chemical causes toxicity, to characterize the abuse liability of a new pharmaceutical,
- determine whether an organism's response to a drug changes with chronic exposure and whether chronic exposure may lead to abuse or physical dependence,
- examine a chemical that is known or hypothesized to have specific

neurobehavioral effects that the investigator wishes to understand in more detail (for example, drugs that block a particular neurotransmitter receptor system can help to determine the neurotransmitter's role in modifying specific behaviors),

- produce a specific neurological state (such as anxiety)
- help researchers to understand the biologic and behavioral consequences and possibilities for therapy. (Weiss and O'Donoghue, 1994.)

## Behavioral and Environmental Considerations

Some neurobehavioral experiments involving drug administration use animals that are trained to perform a response that can be measured objectively. The motivation for the response may be delivery of food or water, or a drug, as in drug self-administration studies (see next section). Trained responses usually involve operating a lever or switch. Other dependent variables may also be measured, such as feeding, drinking, locomotion, or exploratory activity (Iversen and Lattal, 1991; van Haaren, 1993; Wellman and Hoebel, 1997). The research methods reviewed here involve a known substantial risk to humans or animals from exposure to drugs and other chemicals. Additional information about behavioral tests that can be used to screen unknown drugs or genetic mutants is provided in Chapter 9.

Situations requiring special housing or feeding arrangements were summarized in the earlier NIH report (NIH, 2002, p. 58):

> Exposure to drugs usually necessitates individual housing in order to permit repeated access to each animal for dosing and testing. Individual housing also may be preferred because, in a group situation, drug-altered behaviors may increase an animal's risk of abuse by cage mates, as well as impair its ability to compete for food. For animals in studies of intravenous drug self-administration or of constant intragastric infusion, the animal may be fitted with a vest and tether apparatus to protect the chronically indwelling cannula. Behavior may be measured in the animal's living cage, to which devices for presenting stimuli and recording responses have been attached (Ator, 1991; Evans, 1994). Such arrangements may preclude conventional group housing. Experiments in neuropharmacology often employ restricted access to food or water for two purposes: (1) to maintain a consistent motivation of behavioral performance (Ator, 1991) and (2) to standardize content of the digestive tract for uniform absorption and uptake of orally administered drugs. This involves scheduling the availability of food and water but not necessarily deprivation. In addition, for experiments that take place over many weeks, it may be important to keep the total amount of drug delivered relatively constant, even when drug doses are calculated on a per weight basis.

## Dose Considerations

To determine dose-effect relationships, a range of doses is selected—from one that produces little or no effect to one at which significant or even toxic effects are seen. Dose-effect relationships may be determined by studying single doses given to separate groups of animals (between-subject designs) or by determining a full dose-effect relationship for each animal (within-subject, or repeated-treatment designs). Baseline performance usually is reestablished between sessions during which a drug is given. In drug-interaction studies, two doses of different drugs, are given at appropriate intervals before the experimental endpoints are recorded. Cumulative dosing procedures permit increasing doses of a drug to be administered within a relatively short period, and a brief experimental session is conducted after each dose. The effects of the drug are assumed to accumulate in an additive manner so that within a period of 2–3 hours the effects of a range of doses can be determined (Lau et al., 2000; Wenger, 1980).

> Drug self-administration experiments determine the drug's reinforcing efficacy, which may indicate the drug's potential for abuse. The animal controls the number and frequency of delivery of the test drug. That is, a quantity of a particular drug is available intravenously, orally, or via inhalation, and the subject of interest is the amount of behavior maintained by this drug at the self-administered dose. In such studies, the dose available is varied across experimental conditions, and the rate of responding to obtain the dose, the number of drug deliveries obtained, and/or the amount of drug taken are the primary dependent variables of interest. In such studies, the likelihood that the animal will produce a fatal overdose is carefully considered in the design and choice of drug. Drugs vary across classes in how likely it is that high drug doses will produce adverse effects. Overdose may be minimized by placing an upper limit on the number of doses per session or on the minimum time-lapse between doses, or by setting the magnitude of each dose available to the animal (NIH, 2002, p. 59).

## Vehicle Considerations

Drugs for animal research are often in solid form and must be dissolved or suspended in a liquid vehicle to be administered. Sterility of the vehicle is crucial, especially when it is administered intravenously. Aqueous vehicles, such as sterile water and saline solution, have no pharmacologic action of their own in appropriate volumes; however, many drugs require more complex vehicles, for example, such an organic solvent as propylene glycol or an alcohol. Testing with the vehicle, without a drug, will provide a control for the vehicle's influence on the function being studied and for any effects of the drug-administration procedure itself. A vehicle or vehicle-drug combination may irritate tissue. Irritation can be minimized by using less concentrated solutions or alternating injection sites. If less concentrated solutions require volumes that are too large for a single injection site, delivery may be made by small-volume injections at different sites. In some cases, one can adjust the pH to something more similar to physiologic pH by adding another chemical after the drug is dissolved.

## Route Considerations

The route of administration may be dictated by the need to use methods comparable with those of previous neuroscience studies, by constraints on the solubility of the drug, or by a desire to match the route used in humans. The routes of drug administration include oral, subcutaneous, intramuscular, intraperitoneal, intragastric, intravenous, inhalation, or intracranial (for example, into the ventricles or a specific brain region).

Injection by hypodermic needle is the most common way of administering drugs (van Haaren, 1993). The site of injection may be determined by the characteristics of a particular drug's absorption or by the vehicle in which it is given. A common problem is the incorrect site of intraperitoneal injection into rodents. Research staff should be trained to avoid injection into the liver, intestines, or bladder instead of the peritoneal cavity. Success of injections also can be improved by prior adaptation of animals to the handling and restraint that normally accompany injection.

Insertion of a cannula into a blood vessel, a body cavity, or the nervous system is another method of administering drugs. A permanently implanted cannula ensures that repeated injections can be given at precisely the same site and permits the study of drug effects without peripheral effects, such as pain at the injection site (Waszczak et al., 2002). Self-administration studies often use the intravenous route with a chronically indwelling venous cannula (Lukas et al., 1982). The cannula generally is guided subdermally from the intravenous implantation site to exit in the midscapular region. The animal may wear a vest that covers and protects the cannula system. There are also methods for intraventricular drug self-administration through cannulae implanted directly into the brain (Goeders and Smith, 1987).

Implantable pumps for slow drug delivery are also used for chronic delivery, as in studies of drug tolerance or physical dependence (Tyle, 1988). Aseptic technique is important in the implantation of cannulae or pumps and in assessing the system (for example, to reattach tubing or add drug solution), to reduce morbidity and prolong the useful life of the cannulae or pumps.

Inhalation is a common route of exposure for such agents as cocaine, anesthetics, and smoke (e.g., Carroll et al., 1990). Some compounds are easily administered in nasal sprays, but inhalation exposures usually require specialized equipment to measure the amount of drug exposure and to prevent leakage of the airborne chemical (Liu and Weiss, 2002; Paule et al., 1992; Taylor and Evans,

1985). The risk of hypoxia requires attention when drugs are administered by inhalation for long durations.

Oral administration can be used for drug self-administration research (Meisch and Lemaire, 1993). A specialized drinking spout (often termed a drinkometer) regulates the volume of each drink to control drug dose. That permits study of the drug's reinforcing efficacy. Acquisition of taste aversion is studied with oral administration of a toxic substance that serves as an unconditioned stimulus to produce illness. Oral administration of the toxicant can be controlled with a surgically implanted intragastric cannula (Touzani and Sclafani, 2002).

Oral administration is advantageous for chronically administered drugs because dosing may be accomplished without daily handling and intubation if the compound is added to the animal's food or drinking water, as in studies of alcohol self-administration (Cunningham and Niehus, 1997) and exposure to toxic contaminants in food and water (Carpenter et al., 2002; von Linstow Roloff et al., 2002; Weiss and O'Donoghue, 1994). Special feeders and water canisters (Evans et al., 1986) are available to prevent spilling. When a drug is added to food or water, ingestion should be monitored both to determine the amount of drug consumed and to identify reduction in ingestion resulting from reduced palatability. If chronic drug exposure reduces consumption of the food, a control group (for example, pair-fed or pair-watered controls, or in studies done prior to weaning, controls that have restricted access to the lactating dam) should be used to determine whether results are attributable to the drug or to the reduced caloric or fluid intake. Drugs also can be given orally by gavage needle (for example, in rats and pigeons), by nasogastric tube (in monkeys), or in a gelatinous capsule (in monkeys).

### Animal Care and Use Concerns Associated with Toxicity or Long-Lasting Drug Effects

Some chronic drug experiments involve dosing that produces cumulative deleterious effects. The animal-use protocol should include a contingency plan to define the conditions under which deleterious effects will be alleviated or an animal will be removed from the experiment. Some drugs may have long-lasting effects on feeding and drinking, on activity level, and on bodily functions such as elimination. However, other causes of behavioral changes during a drug study, such as irritation at an injection site or dental problems that affect food consumption, must also be examined.

In neuroscience experiments involving chronic drug exposure—for example, to study possible deterioration of performance after repeated exposure to a neurotoxin or the development of tolerance of an initial effect of a drug—attention must be given to the duration of drug exposure and the disposition of the animals. The decision to end chronic drug exposure should be based on predetermined criteria related to a range of changes from baseline that will be considered significant. The observation of overt signs of toxicity, however, may necessitate a decision to terminate treatment earlier than expected. Daily observation of animals by someone familiar with the experimental protocol is especially important so that timely decision-making can occur.

Many dosing regimens do not produce long-term effects or behavioral impairment. After an appropriate washout time, the neuroscientist can determine the existence of long-lasting or irreversible effects (Bushnell et al., 1991). Irreversible effects do not pose a problem if the animal use-protocol calls for the animal to be euthanized to obtain cellular data to supplement functional results. A factor in the decision to euthanize is whether drug exposure has permanently altered a physiologic or behavioral function in such a way as to make providing adequate care for the animal difficult or to compromise continued humane use of the animal. But such an animal would be a valuable resource if the aim of the research is to understand mechanisms of tolerance, postexposure recovery, or therapeutic interventions that ameliorate long-lasting drug effects.

## ADDICTIVE AGENTS

The previous section addressed a wide array of issues related to acute and chronic effects of various chemical agents, including drugs. This section extends that discussion by focusing on issues related to the testing of drugs that are of interest because their chronic use or exposure produces neuroadaptations thought to underlie the behavior patterns (such as tolerance and sensitization, dependence, and withdrawal) that characterize addiction to alcohol, nicotine, cocaine, heroin, and other abused drugs. Neuroscientists study the brain mechanisms that establish and maintain addiction in order to identify and characterize variables that affect risk (for example, genotype, environment, and experience) and to develop methods for treating addictive behavior and preventing relapse (e.g., Koob and Le Moal, 2001). Neuroscientists are also interested in characterizing the neurobiologic consequences of chronic exposure to addictive agents (such as changes in brain structure or function) (Becker, 1996; Obernier et al., 2002) and the process of recovery from deficits induced by such exposure.

### General Considerations

Studies of addictive agents often require attention to dose, route of administration, vehicle, and other variables discussed in the previous section (see also NIH, 2002). When drugs are to be administered with abuse potential, the possibility that an animal will receive a harmful overdose must be carefully considered in the determination of the amount of each dose, the minimum interval between doses, and the total number of doses per session. Those factors depend on drug class, animal species, and, in the case of rodents, strain. They can also vary among individual animals as a function of history of drug exposure, such environmental variables as ambient temperature (Finn et al., 1989), and the presence of a stimulus previously paired with drug exposure (Siegel et al., 1982).

## Animal Care and Use Concerns Associated with Chronic Exposure to Addictive Agents

Studies of the effects of chronic exposure to addictive agents may involve prolonged or repeated exposure to high drug doses over a period of several days, weeks, months, or years. Such studies raise several issues that require consideration. One basic concern is whether extended periods of intoxication interfere substantially with normal feeding, drinking, and other activities (such as grooming) that are important for maintaining the health and well-being of animals. When that concern arises, consideration should be given to alternative methods of providing adequate nutrients and fluids, and of avoiding unsanitary cage conditions.

An additional concern in chronic studies is the possibility that long-term drug exposure will produce long-lasting tissue or functional changes that have adverse effects. In some cases, producing such changes is important to the scientific goals of the study, for example, a study designed to model neurologic deficits associated with Wernicke-Korsakoff syndrome.

In protocols involving prolonged or repeated drug exposure, criteria should be established for determining the duration of exposure and, if necessary, for terminating drug treatment earlier than planned. Daily observation of animals by someone familiar with the experimental protocol is important in such studies to ensure that decision-making is timely.

## Physical Dependence and Withdrawal

In some studies of addictive agents, repeated or chronic drug exposure may produce physical dependence. Physical dependence is revealed by a characteristic withdrawal syndrome on termination of the drug regimen. The salient features and course of the withdrawal syndrome depend on the drug class, the animal species, and, in rodents, the strain (Metten and Crabbe, 1996; Way, 1993; Yutrzenka and Patrick, 1992). And the severity of withdrawal typically depends on the dosing regimen. Withdrawal signs may include irritability, activity changes, body-temperature changes, weight loss, tremor, and convulsive seizures. Drug withdrawal typically produces dysphoria and distress in humans (Jaffe, 1992), and investigators should consider the possibility that withdrawal may produce discomfort and distress in animals.

Whether or how withdrawal is treated in the laboratory will depend on the purpose of the experiment and the nature and extent of the withdrawal syndrome.

In some cases, induction of withdrawal is part of the experimental design, and treatment of the syndrome (for example, with a pharmacologic agent) would interfere with achieving the scientific goals of the study. Nevertheless, even when the schedule of exposure to an addictive agent is designed to allow the expression of a withdrawal syndrome, consideration should be given to establishing contingencies in the event of life-threatening signs, such as excessive weight loss or protracted seizure. Such contingencies might involve supplementary administration of food or fluids through a feeding tube or treatment with an appropriate anticonvulsant drug. When withdrawal is not the subject of the study and the withdrawal syndrome is expected to be severe, dose titration or other drugs may be used to alleviate withdrawal symptoms.

### Occupational Health and Safety Considerations

Use of drugs that are restricted by the US Drug Enforcement Agency (DEA) requires supervision and inventorying by an institutional staff member who is licensed by DEA. The Public Health Security and Bioterrorism Preparedness Act of 2002 and the USA Patriot Act of 2001 require that research institutions collect information regarding hazardous substances classified as "select agents" and register their presence with the federal government.

Staff working with drugs and toxicants must be trained in the use of gloves, gowns, goggles, and eyewash and the appropriate disposal of "sharps." Animals exposed to hazardous materials, including carcinogens and radioactive agents, must be handled and disposed of separately from other animals. Care must also be taken when cleaning the cages or enclosures of these animals to avoid contacting hazardous materials that may have been excreted in the urine or feces. Some hazardous materials may be administered to animals in drinking water from special spillproof canisters to avoid spilling of hazardous materials and exposure of staff to hazardous materials (Evans et al., 1986). If an animal is a possible source of contamination, behavior and physiologic measures can be monitored while the animal remains in its home cage without requiring staff to touch it. Home-cage observations may use a rating scale for cage-side observation, photobeam equipment for detecting locomotion, or telemetry devices implanted in the animal at the start of the experiment, which can be monitored with remote equipment.

## PHYSICAL AGENTS

A variety of physical agents influence neural function, including heat, light, and sound. This section surveys exposing animals to heat, light, and sound to study environmentally induced stress or neural dysfunction.

Some studies of temperature regulation allow animals to indicate a preferred ambient temperature by changing their location. For example, an animal may be given access to a thermal gradient. Animals can also learn to control their ambient temperature with conditioned responses that increase or decrease the ambient temperature of their immediate environment (Carlisle and Stock, 1993; Gordon et al., 1998; Zhong et al., 1996). In other studies, animals may be exposed to inescapably cold or warm environmental conditions (Mechan et al., 2001); in these cases, consideration should be given to providing a period of adaptation to the new temperature, for example, by exposing the animals for increasing periods, or by gradually increasing or decreasing the ambient temperature. In the case of cold exposure, increased availability of food is important.

Changes in luminance and in daily light cycles are used to alter circadian rhythms (Cheng et al., 2002). Such studies may be performed to investigate health problems caused by disturbances in circadian rhythms resulting from jet lag or by shift work.

Auditory stimuli may be studied for their aversive or damaging properties. An important problem in contemporary society is the risk that work-related or environmental noise may damage auditory organs or interfere with auditory perception (Fechter, 1995). Studies of the neurobiology of sensory function or learning may use auditory-reflex methods. With rodents, a brief auditory stimulus is often used to induce a startle reflex (Le Pen and Moreau, 2002). The startle response provides a basis on which to evaluate variables that influence auditory learning and perception. Auditory startle-reflex techniques also are used to evaluate effects of drugs and toxicants that may alter sensory function or response to unexpected stimuli (Crofton, 1992). Other studies involve the use of noise exposure as a general stressor or to cause hearing loss (Fredelius, 1988; Hamernik and Qiu, 2001; Rao et al., 2001). The period and magnitude of noise exposure should be minimized but kept consistent with experimental goals. If noise is not used as an experimental manipulation to produce stress, researchers should recognize that noise may cause stress or induce seizures (Neumann and Collins, 1991). Consideration should also be given to avoiding inadvertent exposure of personnel and other people to excessive noise.

## MODIFICATION OF DIETARY NUTRIENTS

A large body of research focuses on the effects of specific nutrients on neurologic function and dysfunction. For instance, folic acid deficiency in pregnant women leads to neural-tube defects in their children (Werler et al., 1993), and vitamin A deficiency can cause blindness (Anonymous, 1966). When neuroscience and behavioral research involves selective nutrient deficiency or toxicity, the research and veterinary staff must be prepared to deal effectively with the pain and/or distress that may result.

Numerous references identify the clinical signs and physiologic effects of specific nutrient deficiency or toxicity in rodents (NRC, 1995), nonhuman primates (NRC, 2003b), cats and dogs (Aiello, 1998a; NRC, 1985, 1986), and rabbits (NRC, 1977).

An animal use-protocol that involves purposeful deficiency or toxicity of a nutrient should include a comprehensive plan for monitoring the expected physiologic effects. The plan should outline the clinical signs or testing regimen for identifying animals in pain and/or distress, animals at risk of reduced feeding, and animals with physiologic impairments due to the nutritional modification. The animal-care staff should be made aware of the plan, because they may be the first to notice expected or unexpected adverse effects of nutritional modification. Steps should be established in the animal-use protocol and approved by the IACUC in advance to manage the animals adequately without compromising the goals of the experiment and to define clear endpoints for removal of animals from the study.

Many diets used in nutritional studies are ordered, stored, and dispensed outside the normal husbandry operation, so quality control must be ensured. The diets are often administered by the research staff. Record keeping that can be accessed by the husbandry staff and by the IACUC during its semiannual inspection of facilities and animal-study areas is necessary to ensure that animals are being fed in the manner described in the approved animal-use protocol.

## EXERCISE

### Running

The running wheel has been a fundamental tool in neurobehavioral research in rodents since the pioneering studies of Richter (1967, 1971). Voluntary wheel-running is studied to understand neurologic mechanisms controlling circadian rhythms, metabolism, and energy expenditure (Cotman and Berchtold, 2002). If an experiment requires that animals live in the running wheel or in a cage that has been specifically modified to include a running wheel, the cage should comply with the space recommendations of the *Guide*. Forced running, in which a rodent is placed briefly on a moving treadmill or on a rotating bar, is used to measure deleterious effects of drugs on coordination and stamina (see "Behavioral Screening Tests" in Chapter 9).

### Swimming

Rodents are introduced into a pool of water for tests of endurance (Cryan et al., 2002) and learning and memory (Reed et al., 2002). The Morris water maze is one of the most commonly used methods of studying learning and memory in rodents (Barnes et al., 1994; Harker and Whishaw, 2002). Its wide use is based on the rodent's ability to swim without training, and the task requires a shorter training period than do such responses as lever-pressing. Exposure to water at ambient temperature requires less adaptation than exposure to such motivating procedures as food or water restriction. Forced swimming is used to create a standardized stress experience (Griebel et al., 2002; Porsolt et al., 1977). Rodents commonly become immobile after several minutes of swimming if there is no possibility of escape from the water. The dependent variable often is the time until the first episode of immobility or the percentage of the test session spent immobile. Animal-welfare issues include maintenance of an appropriate water temperature, and provision of proper care of the wet rodent after it is removed from the water, and the establishment of unambiguous humane endpoints for testing in the animal-use protocol (see "Mood-Disorder Models" in Chapter 9).

### Animal Care and Use Concerns

Neuroscience studies involving physical conditioning and exercise require appropriate attention to adaptation of the animal to the testing situation, its gradual conditioning to develop stamina, and close animal observation and record keeping during the exercise period. Swim tanks and automated treadmills and running wheels are the most common equipment used to force or promote exercise. Animals should initially be trained on automated devices at low speed, incline, and duration, which should increase gradually as the animals gain stamina. Similarly, the duration of swimming periods should be increased gradually as the animals' condition improves. Weekly or even daily increases may be possible. However, animals should be closely observed by knowledgeable personnel during training and exercise sessions—particularly during the early phases of a conditioning program, near the end of individual training sessions, and during sessions in which performance requirements are increased—and detailed records of the animal's performance and general health should be kept and made available to veterinary staff and the IACUC. In some systems, a rodent's toes or tails may be at risk of becoming entrapped in the treadmill device. The continuous presence of an observer is essential to prevent injury in such situations. The use of remote monitoring systems, such as closed-circuit cameras, is sometimes warranted. As part of its review of the animal-use protocol, the IACUC may consider evaluation of equipment and a preliminary assessment of animal performance in a device.

Many automated treadmill systems apply a mild electric shock to animals whenever they fail to keep up with the programmed pace and drift back on the device. Although the number of shocks experienced by well-trained and conditioned animals is typically low, monitoring and recording shocks that animals experience and the pattern of shock administration during a training session can provide information about the adequacy of the training or exercise in light of the animals' physical condition. A humane endpoint for removal of animals from the testing situation should be specified in the animal-use protocol and approved by the IACUC.

Studies that require an animal to exercise to exhaustion require special consideration. The need for such extreme effort by an animal should be carefully defined and justified, and endpoints should be clearly established and well defined in the animal-use protocol. Specific behaviors, circumstances, or physiological markers should be established to alert the observer that a trial must be terminated. Continuous animal observation is essential near the time of the expected development of animal exhaustion. In all cases, accurate records of test conditions and of performance should be maintained for each animal, and they should be available to veterinary staff and the IACUC. Such records will allow day-to-day adjustment of testing, if warranted by an animal's condition or ability.

A final consideration is the need to maintain sanitation of devices used for exercise or learning. Devices should be constructed of an impervious material to the greatest extent possible. A regular sanitation procedure and schedule should be established, maintained, and clearly documented. A thorough description of the sanitation process should be included in the animal-use protocol.

## SLEEP DEPRIVATION

Short-term sleep loss in humans typically has no adverse physiologic consequences other than increasing sleepiness and impaired performance in some tasks (Horne, 1985; Naitoh et al., 1990). Because sleep is a homeostatic process, adverse effects associated with short-term sleep loss are probably alleviated simply by providing the opportunity to "catch up" on sleep (Everson, 1997; Everson et al., 1989), much as thirst is immediately relieved by taking a drink of water. In rats, biologically significant adverse effects of sleep deprivation have been reported only after sleep deprivation of more than 5 days (Everson and Toth, 2000).

Several approaches are used to produce sleep deprivation in laboratory animals. The method most widely used is probably the so-called "gentle-handling" technique. This method has been applied to rodents, rabbits, and cats and is usually used to cause loss of both rapid-eye-movement sleep (REMS) and non-rapid-eye-movement sleep (NREMS). The animals are under continuous observation by the experimenter and are physically roused by the experimenter whenever they either enter a state of electroencephalographically defined sleep or assume a sleeplike posture. Animals are generally aroused by such actions as tapping on the cage, providing novel objects, or prodding gently. As the duration of the deprivation period increases, particularly beyond a few hours during the species' normal "rest" period, the experimenter must gradually increase the intensity or frequency of handling or of environmental stimulation to maintain arousal (of both the animal and the experimenter!). Because of its labor-intensive nature, the gentle-handling method of achieving sleep loss is rarely extended beyond a 24-hour period. Such short-term sleep loss does not appear to have marked adverse effects in humans or animals other than the progressive development of moderate to severe sleepiness, cognitive and performance impairment, and perhaps irritability or aggression (Everson, 1997; Horne, 1985; Naitoh et al., 1990).

Another relatively common approach to inducing sleep loss in animals is the so-called "flowerpot" technique. This approach produces REMS deprivation by taking advantage of the muscle atony that develops during REMS (Cohen and Dement, 1965; Jouvet et al., 1964). The animals (typically rats) are placed on a small platform (historically an inverted flowerpot) in a tank of water. The platform is large enough to allow the rat to engage in slow-wave sleep, in which residual muscle tone allows it to retain a stable sleeping posture. However, as the animal enters REMS and develops skeletal-muscle atony, it slips from the platform into the water and awakens.

A third approach to causing sleep loss in animals is called the "disk-over-water" technique; it can be used to deprive animals of REMS alone or of both NREMS and REMS (Bergmann et al., 1989; Rechtschaffen et al., 1983). The animals are housed on a rotating platform, or disk, that is positioned over a pan of water. When the electroencephalogram indicates that an animal is entering a state of sleep, a computer algorithm causes the disk to rotate at a low speed. The animal then generally awakens and walks to avoid contacting the water.

A fourth approach is forced locomotion, usually in a slowly revolving drum (Frank et al., 1998; Mistlberger et al., 1987; Rechtschaffen et al., 1999). Animal-welfare considerations relevant to this method are similar to those mentioned previously for exercise models. Interpretation of data collected with this method is confounded by the effect of continuous locomotion or exercise as opposed to the effects of sleep loss itself (Rechtschaffen et al., 1999).

In contrast with the gentle-handling method, the flowerpot and disk-over-water techniques can be easily imposed for long periods, and these approaches create some animal-use concerns. The flowerpot method of REMS deprivation causes alterations in several biochemical indexes of stress (Suchecki et al., 2002). In a refinement of the flowerpot and disk-over-water methods, multiple platforms are used in one large pool so that animals can engage in locomotor activity (Suchecki et al., 2002). Several animals can be housed together in these conditions. Social interactions may reduce some of the nonspecific stress associated with the environment and the physiologic challenge (Suchecki et al., 2002).

Sleep deprivation of over 7 days with the disk-over-water system results in the development of ulcerative skin lesions, hyperphagia, loss of body mass, hypothermia, and eventually septicemia and death in rats (Everson, 1995; Rechtschaffen et al., 1983). The duration of sleep deprivation must be well justified scientifically, particularly if it will be continuous for more than a few days. However, relatively few studies have imposed sleep deprivation long enough to cause those signs. In general, animals that are maintained on chronic sleep-deprivation schedules should be closely monitored for injury and general well-being, and observations should be recorded. The task is simplified by the fact that research teams typically monitor such animals very closely to ensure that they are experiencing the targeted amount of sleep loss. The use of interventions, particularly in chronic studies, must be compatible with the scientific goals of the experiment.

The use of automated sleep-deprivation devices, like the use of exercise devices, requires regular sanitation, good animal observation, and accurate record keeping.

# 9

# Behavioral Studies

The behavior of living organisms is a visible manifestation of activity of the central nervous system. Thus, the study of behavior is a central feature of contemporary neuroscience research in animals. In some studies, the research emphasizes behavior itself, and the primary goal is to characterize behavior and its environmental determinants. In others, the behavior of an animal may be correlated with measurement of brain electric or chemical activity to understand brain mechanisms underlying behavior. Behavioral measures are also used often to detect or measure changes in brain function that may be produced by disease, neural injury, genetic modification, or exposure to various agents and treatments.

The purposes of this chapter are to address several general issues that arise in behavioral studies and to give more detailed consideration to a few specific aspects of neuroscience research in which the measurement of behavior is a central feature.

## USE OF APPETITIVE AND AVERSIVE STIMULI

### Terminology

Stimuli that can be labeled appetitive (attractive or pleasant) or aversive (noxious or unpleasant) are often used in behavioral research. Such stimuli may include food pellets, sweet or bitter tasting solutions, loud noises, drugs, or electric shock. Because use of such stimuli, especially aversive stimuli, is sometimes a source of concern in behavioral studies, this section begins with a brief discussion of the ways in which behavioral neuroscientists commonly describe and categorize these events. In general, appetitive stimuli are ones that an organism will voluntarily make contact with or approach, and aversive stimuli are ones that an organism will try to escape or avoid. Central to those definitions is the idea that the labeling of a stimulus as appetitive or aversive is based on an organism's behavior, not on physical features of the stimuli themselves. Indeed, the same stimulus may be appetitive in some situations or to some individuals, but aversive in other situations or to other individuals. For example, in certain behavioral procedures, rats and monkeys have been shown to engage repeatedly in behaviors that produce exposure to electric shock, an event commonly assumed to be an aversive (Brown and Cunningham, 1981; Cunningham and Niehus, 1997; Cunningham et al., 1993; Kelleher and Morse, 1968). Thus, under these experimental conditions, electric shock would be labeled an appetitive stimulus, not an aversive one. Similarly counterintuitive examples can be found in the literature on behavioral effects of abused drugs. For example, the same dose of alcohol that produces a conditioned place aversion in rats will produce a conditioned place preference in mice (Cunningham et al., 1993). Moreover, a drug's ability to produce a conditioned preference may be completely reversed (to conditioned aversion) simply by changing the temporal relationship between drug injection and the associated stimulus (e.g., Cunningham et al., 1997; Fudala and Iwamoto, 1990). It has also been shown that injection of an abused drug may concurrently induce preference for a paired spatial location, but aversion for a paired flavor solution in the same animal (e.g., Reicher and Holman, 1977). All of these examples illustrate that decisions about whether a given stimulus should be considered appetitive or aversive cannot be based solely on its physical properties, but must be informed by expert knowledge of its behavioral effects in various contexts. Importantly, those effects may vary significantly as a function of the species, genotype, sex, and past experience of each animal.

In more technical terms, the stimuli under consideration here are often referred to as either reinforcers or punishers, depending on their effects in behavioral procedures in which the response-contingent presentation or removal of a stimulus produces either increase or decrease in the rate of the target response. Stimuli that increase the rate of a contingent behavior are called reinforcers, whereas events that decrease the rate of a contingent behavior are called punishers. Both reinforcement and punishment may involve either the presentation or the removal of a stimulus (Domjan, 1998). Typically, the response-contingent presentation of an appetitive stimulus produces an increase in responding (positive reinforcement) and the response-contingent presentation of an aversive stimulus produces a decrease in responding (positive punishment). In contrast, the response-contingent removal or omission of an aversive stimulus produces an increase in responding (reinforcement based on escape or avoidance), whereas the response-contingent removal or omission of an appetitive stimulus produces a decrease in responding (punishment based on omission training).

Although the foregoing definitions indicate that evaluation of behavioral changes produced by response-contingent delivery or removal of such events is critical for applying the labels, the manner in which the stimuli are used in experiments may or may not involve a response contingency. For example, in studies that use instrumental learning or operant conditioning, there will be an explicit, experimenter-defined relationship between some feature of the animal's behavior (such as whether a lever is pressed) and the delivery (or withholding) of the stimulus. In contrast, studies that use classical or Pavlovian conditioning typically do not involve a response-outcome contingency; rather, the emphasis is typically on the predictive relationship between some other stimulus (such as a light or tone) and delivery of the appetitive or aversive stimulus (Rescorla, 1988). Because brain mechanisms underlying the different types of learning may differ, the decision to present appetitive and aversive events in a response-contingent or response-noncontingent manner should be based on the scientific goals of the study.

## Rationales for Using Appetitive and Aversive Stimuli

Appetitive or aversive stimuli are typically used to motivate an animal to perform a particular behavior. However, the scientific reasons for producing that behavior can vary widely, and the overall purpose of the study will be an important consideration in the selection of the appetitive or aversive stimuli. For example, a considerable body of neuroscience research using appetitive and aversive stimuli focuses on understanding the neurobiology of basic motivational processes, such as those involved in feeding, drinking, foraging, mating, drug addiction, aggression, fear (anxiety and phobias), and the avoidance of pain or discomfort. In such cases, the selection of a particular motivational stimulus (such as salt water or a sexually receptive conspecific) is typically dictated by the specific motivational or behavioral system under study (such as sodium appetite or copulation). In other cases, however, investigators may have more leeway in the selection of motivational stimuli. For example, investigators interested in the general neural mechanisms underlying simple learning (such as classical and operant conditioning), cognition, or memory may be able to use a range of stimuli, both appetitive and aversive, to achieve their scientific aims. Similarly, investigators who simply wish to establish a reliable behavioral baseline for studying motor, sensory, attentional, or perceptual processes or for assessing the effect of various manipulations may also have some flexibility in their choice of motivational stimuli. Relevant considerations might include whether the stimulus has similar effects among species or among individuals within a species. Another consideration is the degree of variability in the efficacy of the stimulus among individuals or of repeated exposures to the stimulus in the same individual. For example, because of rapid satiation, a food rich in calories will be a poor choice as a reinforcer in a procedure that requires the animal to respond repeatedly for food over a period of several hours. Of course, the choice of motivational stimuli in such experiments will also be guided by appropriate consideration of their potential to cause pain or distress. This issue is addressed in more detail in the next section.

## Animal Care and Use Concerns

As in other types of research with laboratory animals, investigators conducting behavioral research must consider the recommendations in the *Guide* when making decisions about the choice of appetitive and aversive stimuli. In particular, consideration must be given to Principle IV of the US Government Principles (IRAC, 1985):

> Proper use of animals, including the avoidance or minimization of discomfort, distress and pain when consistent with sound scientific practices, is imperative. Unless the contrary is established, investigators should consider that procedures that cause pain or distress in human beings may cause pain or distress in other animals.

Neuroscientists proposing to use appetitive or aversive stimuli should provide a clear and complete description of the characteristics of the stimulation (such as unit amount, concentration or intensity, duration, and total number) and a scientific rationale for their use in their animal-use protocols. Due consideration must be given to the immediate consequences of acute exposure to these stimuli (for example, do they cause more than momentary or slight pain or distress?) and to possible detrimental effects of long-term or repeated exposure (for example, development of dental caries after prolonged exposure to sugared foods or fluids). Consideration must also be given to possible adverse consequences of restricted access to food or fluids that may be required to provide an appropriate motivational state for the appetitive stimulus (see Chapter 3). As noted earlier, selection of the specific motivational stimuli in a task may be influenced by limitations imposed by the recording techniques; for example, an event that produces little or no movement in an animal may be preferred in sensitive physiologic recording procedures. In some situations, choice of a motivational stimulus and its characteristics will be guided by previous research showing that variability in response to it is low, thus reducing the number of animals that must be used in the procedure. It is also possible that the characteristics of the event must be adjusted individually for each animal to maximize its efficacy or minimize its detrimental effects.

General strategies used by the IACUC, veterinary staff, and members of the research team to evaluate the choice of appetitive and aversive stimuli should mirror those described in previous chapters. In the case of aversive/punishing stimuli with the potential to cause pain and distress, the evaluation process described in Chapter 2 ("Pain and Distress") can be used. As noted earlier, generally acceptable levels of noxious stimulation are those that are well tolerated and do not result in maladaptive behaviors. Use of aversive stimuli at intensities or durations that approach or exceed the animal's pain tolerance level should generally be avoided in behavioral procedures, unless a scientific justification is provided. As discussed previously, it is important to note that the appearance of escape and avoidance behaviors may occur well before the intensity of a stimulus reaches the pain tolerance level. In such cases, these behaviors would be considered appropriate adaptive responses. It is only when the animal's behavior is dominated by escape-avoidance attempts that the behavior becomes maladaptive, signaling unacceptable levels of pain (NRC, 1992).

At first glance, one might assume that avoidance or minimization of discomfort, distress, and pain is more problematic when aversive stimuli are used to motivate behavior than when appetitive stimuli are used. However, that is not necessarily true, especially when one considers that the efficacy of some appetitive foods and fluids depends on the introduction of a restricted schedule of access to food or water (see Chapter 3). Thus, in some situations, an aversive stimulus that does not require prior induction of a "need" state (such as contact with mild electric shock or placement in a pool of water) may actually produce less overall discomfort and distress than the combination of an appetitive stimulus with food or fluid restriction. At the same time, however, one must recognize that detection and measurement of "distress" in animals remains problematic (NRC, 2000) (see "Pain and Distress" in Chapter 2).

In some cases, an investigator's choice of a particular appetitive or aversive stimulus will be determined by scientific reasons. For example, the choice of aversive stimulation such as exposure to electric shock or a predator could be justified by a specific scientific interest in understanding the brain mechanisms underlying behaviors motivated by fear or anxiety. In other situations, however, the scientific question may not directly dictate the choice of one type of stimulus over another. For example, the scientific goals of investigators interested in the neural bases of learning and memory or the mechanisms underlying a specific type of motor behavior might be accomplished by using a broad range of appetitive or aversive events. In situations where the scientific rationale for the choice of a particular motivational stimulus is not compelling or the IACUC is unsure whether one stimulus produces more or less overall discomfort or distress than another (e.g., mild electric shock versus a food pellet combined with food restriction), a useful strategy may be to allow the research to begin using the investigator's preferred stimulus, but to agree in advance to a joint plan for rigorous monitoring and periodic re-evaluation by the IACUC. If apparent pain or distress is higher than expected or other adverse consequences are noted, stimulus parameters can be refined or the stimulus choice changed with approval by the IACUC. If no problems arise during the monitoring phase, the protocol may continue as originally proposed.

An example of the issues that must be considered when evaluating the selection of a behavioral task can be provided by comparing the features of three different procedures commonly used to study spatial learning and memory in rodents: the radial arm maze (Olton and Samuelson, 1976), the Morris water maze (Morris, 1981), and the Barnes circular platform maze (Barnes, 1979). A growing interest in understanding the cognitive decline that accompanies aging together with the recent increase in the number of mouse models carrying genetic mutations thought to affect brain function has encouraged many investigators to use one or more of these tasks to assess "cognitive" function. Although many of the behavioral and brain mechanisms involved in solving these tasks are thought to overlap, the motivational basis for performance differs significantly in each task. For example, the radial arm maze procedure usually involves food or water restriction to motivate animals to seek reinforcers placed at the end of each maze arm, requiring the IACUC to consider the issues discussed previously in Chapter 3 ("Food and Fluid Regulation").

In contrast, the Morris water maze involves immersion in water at or a few degrees above room temperature to motivate animals to swim to a hidden or visible platform. Because exposure to water has the potential for evoking a stress response, the time an animal is in the water should be minimized. Moreover, consideration must be given to drying the animal and providing access to an appropriate heat source (unregulated heating pads and heat lamps should be avoided as they can develop hot spots and cause thermal burns) after water exposure to prevent hypothermia. In the Barnes circular platform maze, the animal is typically placed on a large open platform in a well-illuminated room. The behavior of finding the hole that leads to the darkened escape tunnel located beneath the platform is presumably motivated by the animal's natural aversion to bright open spaces. Some investigators have suggested that this task produces less stress in rats than tasks involving water immersion or food restriction (e.g., McLay et al., 1998). However, the procedures used in several recent studies suggest that additional aversive stimulation (e.g., intense lights, loud sounds, air stream from an overhead fan) may be required to adequately motivate mice to perform in the Barnes maze (Pompl et al., 1999; Inman-Wood et al., 2000; King and Arendash, 2002; Zhang et al., 2002). Thus, at least in mice, this task has the potential to evoke a stress response that may be similar to or greater than that evoked by the other two tasks.

In the case of lesioned or genetically modified animals, the choice of task may be further complicated by sensory-motor impairments that could increase the likelihood of stress or serious injury (e.g., drowning in the water maze, falling off the edge of a Barnes maze). As suggested above, when there is uncertainty about which task produces the least amount of discomfort or distress while still meeting the investigator's scientific goals, the IACUC's best strategy may be to work with the investigator to develop a thorough plan for monitoring the impact of the procedure in conjunction with frequent re-evaluations by the IACUC until the consequences of the procedure are better understood and shown to be acceptable.

## BEHAVIORAL SCREENING TESTS

Behavioral screening tests are used in pharmacology, genetics, and health surveillance (health surveillance through evaluation of animal behavior is discussed in "Using Animal Behavior to Monitor Animal Health" in Chapter 2). Behavioral screening tests differ from hypothesis-driven experiments in that screening tests assess multiple behavioral measures because there is little information to indicate what important effects might be observed. Screening also is used if limitation of time and resources require a test that can be administered quickly to a number of animals. Screening tests are usually directed at broad functional domains, such as motor coordination, emotion, or sensory functions.

Neurobehavioral screens were developed more than 25 years ago to study pharmaceutical and chemical agents (Kulig et al., 1996; Moser, 2000b; Ross et al., 1998). Similar methods are used to screen for genetic mutants (Crabbe et al., 1999b; Sarter et al., 1992a,b; Warburton, 2002).

In reviewing the history of behavioral screening, Warburton (2002) considers the advantages of quantitative methods versus the simplicity of observational screening methods. Observational methods are especially appealing to those with little experience in behavioral science, who may not focus on the possible limitations of observational methods: subjective interpretation, higher variability of baseline behavior, and observational variation among observers. Screening results are most useful if one can demonstrate between-observer reliability, establish standardized protocols, and validate the screen with "gold-standard" procedures.

### Behavioral Screening in Pharmacology and Toxicology

Unlike research protocols for pharmacology and toxicology, drug screening usually implies that the effects of the test compound are not well known. Screening studies can be justified by the need to detect a chemical's ability to cause health problems in humans or animals (such as the abuse liability of a new pharmaceutical or the neurotoxicity of an industrial product) or to determine whether its effects warrant more detailed investigations of its potential as a treatment for behavioral or neurologic disorders. Screening tests also are used when a drug's pharmacokinetics are not well known and observations are required over an unknown time to determine whether an organism's response to the drug changes during chronic exposure and whether such exposure can lead to physical dependence. The IACUC must be aware that regulatory agencies (such as the Environmental Protection Agency or the Food and Drug Administration) sometimes require investigators to use specific test methods and experimental designs (Weisenburger, 2001).

Behavior has proved to be a convenient experimental variable in screening because it is noninvasive and because alterations in many physiologic systems can be reflected in changes in behavior. The functional observational battery (FOB; see Table 9-1) (Moser, 2000a) is a systematic neurologic examination for rodents involving a neurologic examination with numerous behavioral measures. It provides more extensive behavioral measures than the mouse ethogram discussed in Chapter 3 ("Genetically Modified Animals"). The FOB procedure has also been adapted for use with weanling rats (see Table 9-2) (Bushnell et al., 2002; Moser, 2000a). Scoring of the FOB is semiquantitative, and the FOB should be administered and scored by an experienced technician. When a skilled technician is not available, or when handling the animals might be dangerous to animals or staff, observation of an animal in its home cage can be useful, particularly if a quantitative rating scale is used to document the appearance of abnormal behaviors. Better quantification is obtained with commercially available equipment, such as photocell arrays, than through direct observation. The equipment is placed outside a rodent's home cage to measure activity, such as locomotion and rearing, and this avoids the necessity of handling the animal and the possibility that handling may cause changes in the animal's behavior (Evans et al., 1986; Lessenich et al., 2001). Additional methods used in screening for neurotoxicity are reviewed by Weisenburger (2001).

Screening methods for nonhuman primates can be considered along a scale of intrusiveness into the nonhuman primate's living space. Nonintrusive procedures are used to minimize handling the animals. Behavioral activity level, diurnal rhythms, etc., can be monitored with photocell arrays surrounding the home cage (Evans et al., 1989). A food-pickup test can also be used while a primate remains in its home cage. Small pieces of food (such as raisins and peanuts) are systematically placed on a tray and then moved to within the nonhuman primate's reach. The observer measures the time taken to extract the food and the accuracy in terms of the number of attempts required to retrieve all of it (Evans et al., 1989; Merigan et al., 1982). That provides evidence of visuomotor coordination and appetite. Finally, if the experiment permits removing the nonhuman primate from its home cage to a special test apparatus (see "Restraint" in Chapter 3), video cameras can be used to remotely monitor nonhuman primates while they are in the special test apparatus (Ro et al., 1998). A battery of operant conditioning techniques have also been employed to assess neurologic changes caused by a drug or chemical in nonhuman primates (Schulze et al., 1988). This operant screening test is called the Operant Test Battery (OTB; see Table 9-3), and was developed at the National Center for Toxicological Research (NCTR). Additional studies have shown that the OTB can also be used to assess neurological effects in humans and rats (Paule, 2000, 2001).

TABLE 9-1 Functional Observational Battery for Adult Rats

| Endpoint | Measurement/Scale |
|---|---|
| Pupil Response | present/absent |
| Abnormal Body Posture | present/absent |
| Piloerection | present/absent |
| Forelimb and Hindlimb Grip Strength | kg of force |
| Landing Foot Splay | cm/mm |
| Body Weight | g |
| Body Temperature | °C |
| Open-Field Rearing | number |
| Gait Score (also description of gait) | 1 to 5 |
| Ataxia Score | 1 to 5 |
| Aerial Righting Response | 1 to 4 |
| Home-Cage Activity | 1 to 5 |
| Open-Field Activity | 1 to 6 |
| Arousal | 1 to 5 |
| Ease of Removal | 1 to 5 |
| Handling Reactivity | 1 to 4 |
| Tremorigenic Score | 1 to 4 |
| Salivation | 1 to 3 |
| Lacrimation | 1 to 3 |
| Urination/Defecation | 1 to 5 |
| Tail-pinch Response | 1 to 5 |
| Click Response | 1 to 5 |
| Touch Response | 1 to 5 |
| Approach Response | 1 to 5 |

SOURCE: Moser, 2000b.

TABLE 9-2 Functional Observational Battery for Pre- and Post-weanling Rats

| Endpoint | Measurement/Scale |
|---|---|
| Body Weight | g |
| Open-Field Rearing | number |
| Gait Score (and description of gait) | 1 to 3 |
| Forelimb Grabbing | 1 to 4 |
| Surface Righting Response | 1 to 4 |
| Open-Field Activity | 1 to 6 |
| Arousal | 1 to 5 |
| Handling Reactivity | 1 to 4 |
| Tremorigenic Score | 1 to 3 |
| Urination/Defecation | 1 to 5 |
| Lacrimation | 1 to 3 |
| Salivation | 1 to 3 |
| Tail-Pinch Response | 1 to 5 |
| Click Response | 1 to 5 |

SOURCE: Moser, 2000b.

TABLE 9-3 NCTR Operant Test Battery

| Function | Name of Test |
|---|---|
| Motivation | Progressive Ratio Task |
| Discrimination | Color and Position Discrimination Task |
| Timing | Temporal Response Differentiation Task |
| Short-term Memory | Delayed Matching-to-Sample Task |
| Learning | Repeated Acquisition Task |

SOURCE: Paule, 2000.

An important consideration for the IACUC, researcher, and veterinarian is the selection of the animal species to be used for behavioral screening in pharmacology and toxicology (Luft and Bode, 2002). Available data on kinetics and metabolism should be taken into consideration in identifying a species whose behavior will best predict effects in humans. Generally speaking, rodents are good models for behavioral screening in studies of neurotoxicity and neuropharmacology (Luft and Bode, 2002).

Some behavioral-toxicology experiments involve dosing that produces deleterious effects. The protocol should provide a contingency plan for conditions in which side effects will be alleviated or that require an animal's removal from an experiment (see "Animal Care and Use Concerns Associated with Toxicity or Long-lasting Drug Effects" in Chapter 8).

## Behavioral Screening of Genetically Modified Animals

### General Considerations

Once a general health assessment of a newly developed strain of genetically modified animals is completed (see "Genetically Modified Animals" in Chapter 3), behavioral phenotyping should proceed as soon as sufficient numbers of transgenic animals are available to identify sensory, motor, or motivational deficits that may compromise animal well-being. Sensory and motor assessments should be completed before assessment of more complex behaviors—such as learning and memory, aggression, mating, and parental behaviors—because sensory and motor deficits may confound the interpretation of other behavioral assessments.

Behavioral tests assess the effects of altering, adding, or removing a gene (and gene product) on behavior, not the effects of the normal gene on behavior (Nelson, 1997). Behavioral phenotyping can also be confounded by impairments that are secondary to the missing or inserted gene; for example, knocking out a gene may cause the compensatory overexpression of a second gene and any changes in behavior could be the result of the overexpression of the second gene. Those possible problems can be overcome in the same way as in other types of ablation studies: by collecting converging evidence with a variety of pharmacologic, lesion, and genetic manipulations. Because mammalian genome mapping currently focuses on mice (*Mus musculus*), standardized behavioral testing of mice should be adopted (Brown et al., 2000; Crawley, 2000).

Altered behaviors of knockout mice are often sufficiently obvious or unusual that they catch the attention of animal-care personnel, who then notify the investigators. Dramatic behaviors that include increased aggression, altered maternal care, decreased sexual behaviors, seizures, and impaired motor coordination and sensory abilities are commonly reported for knockout mice (e.g., Barlow et al., 1996; Brown et al., 1996; Brown et al., 2000; Chen et al., 1994; Crawley, 2000; Nelson et al., 1995; Saudou et al., 1994). Presumably, knockout mice may have more subtle behavioral changes that have not yet been discovered, even among mutants with no obvious behavioral phenotypes. Some of the behaviors probably will be revealed only if the animals are housed in conditions that are ecologically relevant with respect to space and social organization (Cabib et al., 2000; Pfaff, 2001; Potts et al., 1991).

Behavioral performance is compared among wild-type (+/+), heterozygous (+/–), and homozygous (–/–) mice in which the gene product is produced normally, produced at reduced levels, or missing, respectively. The comparison of +/+ and –/– littermates of an $F_2$ recombinant generation is probably the minimal acceptable control in determining the behavioral effects of knocking out a gene or genes (Morris et al., 1996).

In the past, many knockout strains were generated by using stem cells from one strain and blastocysts from another strain (see "Knockout and Knockin Mutants" in Chapter 3 for review). Therefore, behavioral differences shown by knockout mice may reflect strain effects rather than the effects of the absence of the missing gene (Broadbent et al., 2002; Gerlai, 1996; Threadgill et al., 1995). Given the potentially important effect of background genotype on ability to detect effects of targeted mutations (Crabbe et al., 1999a; Lariviere et al., 2001), behavioral neuroscientists should attend to the genetic background of the transgenic animals under study to ascertain that proper controls for strain differences are used. Another limitation of the interpretation of behavioral data from knockout mice is the possibility that compensatory or redundant mechanisms might be activated when a gene is missing. For example, mice lacking the gene for the neuronal isoform of nitric oxide synthase (nNOS–/–) have a 20% increase in the expression of the endothelial isoform of nitric oxide synthase (Burnett et al., 1997). A compensatory mechanism may spare behavioral function and cloud interpretation of the normal contribution of the gene to behavior. Knockout mice are almost always raised by their natural mothers, which are missing one or more genes that may directly or indirectly affect maternal behavior. Thus, any changes in behavior observed in the knockout offspring may reflect the absence of the missing gene or reflect alterations in maternal care. Cross-fostering of matched-size litters can be used to untangle these influences.

The availability of "inducible" or "conditional" knockouts, in which a specific gene can be inactivated at any point during development or inactivated only in tissue-specific cells, should provide an important tool to bypass the problem of developmental interactions (Holmes, 2001; Nelson, 1997).

### Sensorimotor Tests

Some knockout mice have sensory or perceptual deficiencies that can confound interpretations of altered complex behaviors, such as learning, parenting, mating, or aggression (e.g., (Cases et al., 1995; Young et al., 2002). For example, genetically altered mice may suffer from retinal degeneration and fail to perform adequately in tasks such as use of a Morris water maze, which often requires that animals use extramaze visual cues (Hengemihle et al., 1999).

Vision is assessed with a variety of tests, including the visual-placement test and the visual cliff (Zhong et al., 1996). Auditory abilities are assessed with either a clicker-orientation test or an acoustic-startle test (Crawley et al., 2000; Jero et al., 2001; Weisenburger, 2001). Olfactory ability is determined by how long it takes an animal to discover highly odoriferous food (such as cookies, peanut butter, bacon, or cheese) hidden beneath the cage bedding (Nelson et al., 1995) or by odor-discrimination tests (Sundberg et al., 1982). Pain sensitivity can be tested with paw removal from a hot plate or a tail-flick test (Rubinstein et al., 1996); the motor skills of transgenic mice should be assessed before this test to avoid tissue damage caused by slow reaction, rather than high pain thresholds, but in any case, the stimulus should always be terminated after a predetermined time interval selected to avoid tissue damage. The proposed procedures for assessing general motor skills in transgenic mice before behavioral testing and the criteria for early removal of an animal from a potentially painful or distressful stimulus should be described in detail in the animal-use protocol.

### Motor Tests

After assessment of sensory abilities, motor abilities and coordination should be assessed. Many strains of mice (such as waltzers, weavers, and staggerers) suffer from movement difficulties that could affect locomotion, coordination, or grooming (Brown et al., 2000). Such motor deficiencies could also affect performance in any behavioral assessment that requires movement (such as depressing a lever or running a maze) or performance of specific behaviors (such as aggression, mating, or parenting). Many transgenic mice display movement or gait disorders (for example, dopamine 1A receptor–/– and GM2/GD2 synthase–/– mice). For instance, mice that are engineered to lack a key enzyme in complex ganglioside biosynthesis (GM2/GD2 synthase) and that express only the simple ganglioside molecular species GM3 and GD3 develop substantial and progressive behavioral neuropathies, including deficits in reflexes, strength, coordination, and balance (Chiavegatto et al., 2000). Quantitative tests of motor abilities determined at the ages of 8 and 12 months revealed progressive gait disorders in complex-ganglioside knockout mice compared with controls, including reduced length and width of stride, increased hindpaw print length, and marked reduction in rearing. Compared with controls, null mutant mice tended to walk in small labored movements and performed poorly in many tasks that required coordinated movements (Chiavegatto et al., 2000).

### Assessment of Anxiety

Genetically altered animals may differ from wild-type animals in their emotional responses (fear, anxiety, and defensive reactions). Atypical emotional reactions interfere with responses in learning and memory tasks and with assessment of mating, parenting, or aggressive behaviors.

Several assays of anxiety-like behaviors have been developed. The most common are the so-called exploration-based tests (Holmes, 2001). The premise of these tests is that for some species such as rodents, the innate drive to explore a novel place will be inhibited as aversion to the new space increases. A simple version is the open-field test. High levels of exploration of the open, brightly illuminated area of an enclosure are interpreted as low-anxiety behavior (reviewed by Holmes, 2001). Highly anxious mice stay near the wall of the enclosure. Administration and scoring of this test have been automated, and several commercial products for performing the test are available. Defecation constitutes an additional measure of anxiety; high rates of bolus production are correlated with anxiety in wild-type rodents. Treatment with anxiolytics increases the time spent in the "open" portion of the open field and reduces the number of boluses produced (Holmes, 2001). Obviously, gene manipulations that affect metabolism or food intake could affect bolus production and confound the assay of anxiety. Other exploration-based tests of anxiety include the elevated plus maze, the light–dark exploration test, the emergence test, and the free-exploration test (Belzung and Griebel, 2001; Pare et al., 2001).

The elevated plus maze has become the most commonly used screen for novel anxiolytics, as well as a probe for anxiety in transgenic mice (Holmes, 2001). The elevated plus maze is shaped like a plus sign, has two open arms and two enclosed arms, and is usually raised at least 1 meter above the floor (Lister, 1987). The test animal is placed in the open center of the plus maze, and the number of entries into the closed arms is compared with the number of entries into the open arms over some period (commonly 5–15 minutes). High levels of anxiety correlate with more time spent in the enclosed arms.

The light–dark exploration test is based on rodents' innate preference for small, dark, enclosed spaces over large, light, open spaces and their innate tendency to explore novel environments (Crawley, 1981). Spending more time in the light side of a box than in the dark side indicates low anxiety.

All the exploration-based tests of anxiety rely on movement of an animal in a test apparatus. As noted, many genetically modified animals have motor difficulties. The motor deficits must be taken into account in evaluating anxiety. For example, a common assay for fear is freezing in response to a loud noise (startle response). If a transgenic mouse being tested is slow to move, it may appear to be "freezing" for longer periods than its wild-type cohorts and thus appear to demonstrate a high level of fear. The startle response can be modified by classical conditioning, but it also requires "normal" motor abilities. Only after health problems, sensorimotor deficits, and atypical emotional responses in genetically modified animals are ruled out should behavioral assessment proceed. Rodgers (2001) reviews methodologic pitfalls that should be considered by investigators seeking to characterize genes related to anxiety.

## Animal Care and Use Concerns

Animal-care personnel are likely to discover behavioral deficits (such as lack of feeding or maternal care) and should be trained to recognize them. Neuroscientists should be aware of *potential* problems with animals exposed to novel drugs or neurotoxins and with transgenic animals before their development, housing, breeding, or experimental use. Because of potential sensorimotor deficits or general frailty, transgenic mice should be monitored at least once a day by trained observers until the limitations of the animals are well characterized. That concern is especially appropriate when using tasks—such as those involving water immersion, roto-rods, elevated platforms, treadmills, or other mechanical devices—in which there is a high risk of injury to impaired animals. In animals showing specific sensorimotor deficits caused by a neurotoxin or genetic manipulation, it may be necessary to choose tasks or modify tasks so that they do not impose demands beyond the animal's reduced abilities. For example, the circular platform maze or radial arm maze described earlier will be better choices than a water maze for testing cognitive function in mice with severe motor impairments that interfere with swimming. Because screening procedures often involve testing a given animal in multiple tasks, excellent record keeping practices are imperative. For all the behavioral phenotyping assessments, clear end points (both temporal and performance) for removing animals from the protocol must be identified. Given the unique deficits that may arise from exposure to novel drugs, neurotoxins, or genetic manipulations, it may be necessary to develop different endpoints for the removal of experimental animals than for normal control or wild type animals. Proposed procedures for dealing with each of those issues should be described in detail in the animal-use protocol, and they should be carefully reviewed by the IACUC. As suggested previously, in situations where the consequences of an experimental manipulation are uncertain or unknown, IACUCs are well advised to work with investigators to develop a plan for careful monitoring and periodic re-evaluation to ensure the health and well-being of the animals.

# NEUROPHYSIOLOGIC RECORDING IN AWAKE, BEHAVING ANIMALS

This section discusses some of the issues to consider in preparing and maintaining animals that are used in neurophysiologic recording experiments while they are awake and performing a behavioral task. Technical issues and animal care and use issues are discussed at length in Chapter 4 ("Neurophysiology Studies").

Because animals are awake during experimental sessions, many of the concerns associated with studies on anesthetized animals are avoided. However, specific considerations are warranted in these studies. Researchers invest a considerable amount of time and effort in the behavioral training and surgical preparation of each animal. As noted in *Preparation and Maintenance of Higher Mammals during Neuroscience Experiments* (NIH, 1991), the extensive training and surgical preparation, as well as the often long-term experimental use of the animal, presents a number of issues that require the use of professional judgment and flexibility in interpreting the recommendations of the *Guide*. Adoption of a team approach to these types of studies is essential to ensuring animal well-being and the acquisition of the necessary data.

The study of many important neuroscience questions requires the use of an awake, behaving animal. Behaving preparations make it possible to study cognitive and integrative brain processes by engaging an animal's active participation. Trained animals can serve as subjects in experiments directed toward understanding of the neurophysiological or psychophysical processes that underlie motor control, sensorimotor integration, learning, memory, and perception (NIH, 2002). The behavioral repertoires of many mammals resemble those of humans, so data generated on awake, behaving animals can be expected to have considerable relevance when extrapolated to human behavior and neurophysiology (NIH, 1991). An understanding of the species-typical behavior of the animals used in awake, behaving experiments is critical for adequately assessing the animal for signs of stress/discomfort/frustration that may be minimized either through an earlier endpoint determination or by modifying experimental procedures or study-related equipment. Additionally, such knowledge will assist personnel in avoiding the use of inappropriate visual cues (for example, a direct stare at a macaque or large hand/arm gestures) that the animal might perceive as threatening or stress-inducing.

## Behavioral Training

Experiments on awake, behaving animals generally occur in several stages. Usually, an animal learns to perform a task reliably during an initial training phase. Concurrently with or immediately after this training phase, various devices that are required to quantify behavioral variables, such as eye coils to monitor eye position (Fuchs and Robinson, 1966) or the neurophysiologic consequences of task performance, such as electromyographic electrodes to record muscle activity (Loeb and Gans, 1986) are surgically implanted. (Some of the devices that are typically implanted are described in greater detail below.) After implantation of necessary hardware, data are collected in regularly scheduled recording sessions over a period that can extend for months or years. The use of these animals for this long a period of time makes it even more incumbent upon research personnel to understand the effect of their behavior on that of the animals with which they work. For example, Bayne et al. (1993) demonstrated that positive interaction with nonhuman primates can lead to significant reductions in the expression of abnormal behavior, while Line et al. (1989) have shown that even routine husbandry procedures performed by familiar staff can influence an animal's physiology, such as heart rate. The potential impact animal care staff can have on animal well-being has recently been reviewed (Bayne, 2002). Indeed, it has been recommended that "a genuine caring attitude" prevail among animal-care staff.

Experiments with behaving animals may involve training animals to perform a specific behavioral task. That allows the neuroscientist to repeatedly elicit and monitor a stereotyped movement, to present sensory stimuli under highly controlled conditions, and to obtain psychophysical discriminations from the animals. In addition, providing animals with a challenging and rewarding behavioral task can stimulate their cooperation in the experiment, reduce their boredom, and generally facilitate collection of high-quality data (NIH, 1991). If the goal of an experiment is to examine variables associated with learning, naive animals are often studied as they learn a new behavior (Lemon, 1984c). That allows the behavioral or neurophysiologic variables that change as a skill is acquired or refined to be compared and measured.

Various training methods and tasks are suitable for achieving experimental goals. In most neuroscience experiments on awake, behaving animals, traditional operant-conditioning paradigms are used. These paradigms require an animal to respond behaviorally to a stimulus to achieve a desired consequence. The most common procedures used in neuroscience experiments are appetitive or aversive (see above, "Use of Appetitive and Aversive Stimuli"). Aversive procedures generally involve exposing animals to some form of noxious stimulus (such as mild electric shock, a bitter food, or an unpleasant sound) when they make an incorrect behavioral response. For some experiments, aversive procedures may be the most appropriate means of training animals to perform a task, because they yield highly reliable behavior with smaller differences between individuals (Toth and Gardiner, 2000). Aversive procedures may also be less likely to upset basic metabolic functions than appetitive procedures. However, it is generally recommended that the use of such aversive procedures conform with Principle IV of the US Government Principles (IRAC, 1985), which states, "Unless the contrary is established, investigators should consider that procedures that cause pain and distress in human beings may cause pain or distress in other animals." Thus, the level of aversive stimulation applied to an animal generally should not exceed that tolerated by a human being. Not uncommonly, a member of the IACUC will experience the aversive stimulus for him/herself to better understand what the animal will undergo.

Restriction of access to food or water is often used in behavioral neuroscience and neurophysiologic recording paradigms to motivate animals to execute desired behavioral tasks. This process merits attention to specific considerations that are addressed in Chapter 3 ("Food and Water Regulation"). In brief, investigators should provide a sound rationale for using appetitive or an aversive procedures. If access to liquid or solid food is to be restricted, the proposed level of dietary control should be justified, and appropriate monitoring and record keeping procedures should be described in the animal-use protocol. The procedures can be based on the literature or on an investigator's own experience and should include criteria for determining intervention endpoints for removal of an animal from a particular conditioning paradigm. The goal of the monitoring procedures is not only to keep animals in a highly motivated state but also to maintain their health and welfare. Accordingly, records often include details regarding the animal's performance on the behavioral task and various physiologic indexes.

## Documentation and Record Keeping

Because of the potential health implications of food or fluid restriction, the health status of animals used should be well documented if food or water availability is restricted. Representative animal records might include weight or assessment of hydration status, general appearance or disposition, performance during the behavioral-task session, volume of fluid consumed (earned plus supplemented), dietary supplements or treats that were given, and experimental manipulations that were performed or treatments that were administered. For some species, the welfare of the animal may be further assured by monitoring its behavior in the home cage. Investigators should weigh animals according to suggested guidelines (NIH, 2002) and be alert to changes in mood, behavior, or appearance that indicate a potential medical concern. Individual animals may respond adversely to the weighing process; in such cases, judicious adjustment of weighing frequency or modifying the means of obtaining weights to better accommodate the individual animal may be necessary, and alternative methods of monitoring hydration may be advisable, for example, skin turgor, moistness of feces, and general appearance and demeanor (see "Methods for Assessment of Proper Nutrition and Hydration" in Chapter 3).

There is some likelihood of weight loss during different phases of training (NIH, 2002). An animal's use in a food- or fluid-restricted behavioral experiment should be assessed with veterinary input if persistent weight loss occurs. No single physiologic measurement will always provide a reliable index of an animal's well-being throughout the course of a behavioral neurophysiology experiment, but regular monitoring of several measurements (such as of food and water intake, weight, urine and feces, fur and skin, and behavior) usually permits adequate noninvasive evaluation. Each animal has different needs for food and fluids, so flexible criteria are preferable to rigid prescriptions of how much food or fluid animals should receive daily.

Because different animals respond to food or fluid restriction challenges with different physiologic and behavioral accommodations, monitoring of each animal is essential, and adjustment of the restriction protocol is sometimes necessary (Toth and Gardiner, 2000). That is especially important during the initial stages of learning a new behavioral task. Emphasizing the role of professional judgment in these types of experiments, Toth and Gardiner (2000) recommend that:

> If task performance is not adequately supporting minimal intakes, the experimenter should re-evaluate and perhaps simplify the training strategy to facilitate the animal's ability to learn and master the task.

Standard clinical tests will reveal serious pathologic conditions, but the more insidious, gradual deterioration of an animal's status can be recognized and treated only if there is regular observation and the implementation of professional judgment. Perhaps the greatest challenge in the maintenance of awake, behaving animals is the determination of their overall status. An animal's overall behavior in its cage is a sensitive indicator of its psychologic and physical status (NIH, 1991). Investigators, veterinary personnel, and when available, behavior experts share in the responsibility of observing behavior, general appearance, and demeanor throughout an experimental regimen. Handlers of animals that are used in behavioral experiments should be knowledgeable and skilled in the interpretation of behavior such that changes that could indicate underlying health and well-being problems are readily identified and reported (Bayne, 2000). To this end, each animal should serve as its own behavior control, with baseline observations made prior to the initiation of the study.

## MOOD-DISORDER MODELS

There has been considerable debate about the validity of animal models of human affective disorders. At a minimum, a good animal model of an affective disorder should meet many or all of the following criteria (Redei et al., 2001): strong behavioral similarities with the human disorder, a cause similar to the cause of the human disorder, similar pathophysiology, and similar treatments. Several animal models of affective disorder have been developed, especially for depression. In these models, depressive behavior may be caused by genetic manipulation, environmental perturbations or stressors, or drug treatments (Redei et al., 2001). As is the case in all behavioral research, care must be taken to assess performance in these model systems in a valid and reproducible manner.

## Depression

The so-called Porsolt swim test is the most commonly used test for assessment of depression in animal models (Porsolt, 2000; Porsolt et al., 1977). Other tests and procedures include the tail-suspension test, anhedonia (such as with consumption of sucrose solution), learned helplessness, chronic mild stress, olfactory bulbectomy, differential reinforcement of low rate of responding behavior, and conditioned place preference (Porsolt, 2000; Redei et al., 2001; Vaugeois et al., 1997; Willner, 1997). In all those tests, treatment with antidepressants that are effective in treating humans with depression reverses the depressed behavioral responses. It is generally accepted, however, that the Porsolt swim test (behavioral despair) and tail-suspension tests model human depression most closely (Crawley, 2000; Porsolt, 2000).

In the Porsolt test, rodents are placed in a container of water at least 30 cm deep (to prevent an animal from touching the bottom of the container with their tail) and at least 15 cm from the top of the container (to prevent escape). To avoid temperature-related stress responses, the water temperature should be 24–30°C. Rodents placed in water generally swim, but if manipulated with some drugs or brain lesions, they stop swimming and float. Floating is considered a measure of depression because the animals appear to stop trying (learned helplessness or behavioral despair) and because drugs that are effective antidepressants in humans decrease floating time (Crawley, 1999). Genetically modified animals may require special attention; any rodent that fails to swim or float should be removed from the water immediately. However, even if transgenic animals can remain afloat, locomotor difficulties can interact with performance in the Porsolt swim test and cause nondepressed transgenic animals to appear depressed.

The tail-suspension test avoids the problems of locomotion somewhat and avoids the hypothermia and stress associated with forced swimming (Vaugeois et al., 1997). Animals are suspended by their tails and the amount of "immobility" is measured by a force-strain gauge that records all their movements (Steru et al., 1985). Longer periods of immobility are associated with higher depressive scores. The immobility can be reversed with antidepressant treatment (Vaugeois et al., 1997).

Reduced ingestion of a sucrose solution is another reliable indicator of depression-like-behavior in rodents (Gittos and Papp, 2001; Stock et al., 2000). This test avoids some of the problems of locomotion and coordination of the Porsolt test, but if the targeted gene affects metabolism or food intake, its reliability for depressed behavior may be impaired.

In all those behavioral tests of depression, proposed procedures for monitoring, record keeping, and humane intervention should be described in the associated animal-use protocol and approved by the IACUC.

### Anxiety

Assessment of anxiety was described earlier in this chapter ("Behavioral Screening of Genetically Modified Animals, Assessment of Anxiety").

### Alcohol and Drug Addiction

Several testing paradigms assess responsiveness of rodents to drugs that have abuse potential, including paradigms involving self-administration of alcohol, cocaine, morphine, or nicotine (Crawley, 1999; Grahame and Cunningham, 1995). Self-administration is typically achieved by requiring the rodent to press a lever or display a place preference. Tolerance, dependence, and withdrawal symptoms can be studied. With this approach, transgenic mice may have locomotor or coordination difficulties that interfere with self-administration (McClearn and Vandenbergh, 2000). Additional information is provided in Chapter 8 ("Addictive Agents").

### Animal Care and Use Concerns

The primary goal of the preceding behavioral assays is the induction of stress or aversive states. It is important for the investigator to determine the earliest possible or least severe endpoint when the manipulation has adverse effects on an animal. Any behavioral test that subjects animals to water has the potential for evoking a stress response. Therefore, it is important that the time that the animal is in the water be minimized and that animals be monitored closely to avoid unnecessary stress. Animals should be dried thoroughly after the swim test, and it is advisable to place their cages on a heating pad for several minutes. The use of unregulated heating pads or heat lamps should be avoided as they can develop hot spots and cause thermal burns.

Continuous monitoring is also important for automated tasks, such as tasks that use roto-rods, platforms, or other devices in which animals may be injured. Because of the likelihood of multiple testing, excellent record keeping is imperative.

## BEHAVIORAL STRESSORS

Some neuroscience research involves exposing animals to behavioral stressors. These manipulations can be social (such as involving social separation or mixing of unfamiliar animals) or nonsocial (such as exposing animals to novel environments or restricting behavioral activity).

This research focuses on three avenues of investigation. The first is aimed at understanding the effects of exposure to behavioral stressors on aspects of neural function or conversely understanding how neural manipulation affects responses to behavioral stressors (von Borrell, 1995). For example, a pregnant monkey might be exposed to various behavioral stressors, such as noise and unfamiliar surroundings, and neurochemicals associated with the stress response would be measured in her offspring to determine the effects of prenatal stress on the development of stress responsiveness in young animals (Schneider et al., 1998).

The second is aimed at understanding the neural substrates or correlates of particular behaviors or aspects of temperament, including social recognition, affiliation, pair bonding and attachment, parental behavior, social dominance, aggression, predation, play, and fearfulness (Amaral, 2002; Kavaliers and Choleris, 2001; Siegel et al., 1999; Young, 2002). In those studies, animals may have lesions, be genetically modified (mouse knockouts), or be electrically or chemically stimulated, and the resulting behaviors can be observed; or neural function may be measured during or after the performance of the behaviors of interest.

The third category consists of pharmacologic studies to determine the efficacy of various compounds in reducing aggression, anxiety, or fearfulness (Mench and Shea-Moore, 1995). The purpose of those studies is usually to identify compounds that may be useful in human or veterinary clinical medicine, but pharmacologic testing can also be used for studies of underlying mechanisms of behavior: the behavior of interest is stimulated in some way, usually by staging an aggressive encounter or placing an animal in a fear-inducing situation, and compound efficacy is then evaluated with behavioral measures.

## Social Disruption

Social disruption can be used as an experimental technique in neuroscience and behavioral research, but it can also be an inadvertent confounder of the research. Experimental designs that purposefully incorporate social disruption, do so through the temporary removal and reintroduction of offspring or of group or pair-mates, longer-term or repeated reorganization of social groups by removal of group members or by introduction of unfamiliar animals to groups or to one another, or even the merging of different groups of animals. Abnormal social conditions can also be created by placing animals in atypically small or large social groups, by forming groups of atypical composition (such as all-male groups or groups comprising only animals of similar age), or by crowding them. In addition to the study goals described above, this technique has recently been used to study coronary artery atherosclerosis, heart rate reactivity, and the effects of exercise in conjunction with social disruption on coronary heart disease (Kaplan et al., 1982, 1993; Manuck et al., 1983a,b; Williams et al., 1991, 2003).

The effects of social separation (such as individual housing) or social isolation on an animal's behavioral profile have been documented in various species. The impact of social separation or isolation can depend on the species or strain of animal, the age at which an animal is removed from conspecifics, the duration of the separation, and the completeness of the separation (with respect to visual, auditory, or olfactory cues from other animals). In nonhuman primates, the lack of physical contact appears to be the most important cause of abnormal behavior, both in infants and in adult animals (Bayne and Novak, 1998).

Animals that are isolated to disrupt the infant-parent bond often display acute responses to indicate stress. Distress vocalization, changes in general activity and heart rate, as well as elevated cortisol/corticosterone concentrations can occur and are adaptive under normal circumstances. However, if the separation is prolonged, as during experiments where the effects of infant-parent bond disruption are being studied, it becomes distressful and can lead to maladaptive behaviors as the infant animal matures. Self-injurious behaviors, stereotypic behaviors, extreme timidity or aggressiveness, and inability to mate or provide adequate care to offspring are maladaptive behaviors that might result from the social disruption (NRC, 1992).

Kittens separated from their mothers at an early age tend to be more aggressive and nervous as adults (Seitz, 1959), and social play is critical for a kitten's development (O'Farrell and Neville, 1994). Puppies that are not adequately socialized to other dogs or people may be excessively fearful or aggressive (O'Farrell, 1996). Wolfle (1990) has described a puppy-socialization program and behavioral scoring method specifically for use in the research environment. Monkeys reared in partial or total social isolation develop a syndrome of behavioral abnormalities that includes rocking, huddling, self-clasping, and excessive self-orality (Cross and Harlow, 1965; Harlow and Harlow, 1965). As the animals age, stereotypic patterns emerge, such as repetitive locomotor patterns, floating limbs, and eye poke or salute. The isolation syndrome is also manifested in the development of abnormal social relationships (Mason, 1968).

A restricted social environment can also affect adult animals. For example, long-term (2-year) individual housing of adult nonhuman primates has been shown to alter social behavior (Taylor et al., 1998). Unless the research focuses on social restriction or veterinary concerns develop, infant animals should be reared in a social environment with mother and peers, with mother only, or with peers only to reduce or prevent psychopathologic conditions (Bayne and Novak, 1998). Similarly, when the research, health, and safety of the animals allow it, adult social animals should be maintained in a social environment (for example, pair- or group-housed).

### Animal Care and Use Concerns

The primary animal care and use concern associated with social disruption is the distress that leads to the display of maladaptive behaviors. When studies involve the use of social disruption, the animal-use protocol should include humane endpoints for removal of the animal from the study. Determining endpoints that are predictive of severe distress is a matter of professional judgment and should evolve through discussions between the IACUC, veterinarian, and PI.

It is important to recognize that the display of the maladaptive behavior affects not only the isolated animal but can also have unintended affects on the dam (in studies of infant-parent bonds), the potential offspring of the animal, and the conspecifics that may be forced into the animal's social group. In some species, such as nonhuman primates, dams also show a response to separation from their infants. Their behavioral and physiologic reactions appear to be similar to those of the infant, although less persistent and intense (NRC, 1992), and steps should be taken to minimize this distress if it is an unintended byproduct of the experiment.

The offspring of animals in social disruption experiments may also be impacted by the maladaptive behavior of its dam. For example, female rhesus macaques that are isolate-reared can be neglectful or abusive of their infants (Suomi, 1978). In that situation, it may be appropriate to provide additional support to the offspring or protect it from injury.

In some cases, social disruption causes aggression toward conspecifics. For example, social restriction of male mice will lead to intermale fighting (Brain, 1975). Similar findings have been observed in gerbils, hamsters, and rats (Karim and Arslan, 2000; Payne, 1973; Wechkin and Breuer, 1974). Isolation-reared rhesus monkeys are hyperaggressive and do not develop normal social relationships with other monkeys (Anderson and Mason, 1974; Mason, 1961); this aggression can be directed to other animals or be self-directed (Gluck et al., 1973). Steps should be taken to prevent injury in these cases. For instance in nonhuman primates, this may require housing the aggressive animal separately (AWR 3.81(a)(1)) or the use of screen barriers within cages to permit side-by-side contact, but prevent agonistic encounters.

## Induced Aggression or Predation

Several common models are used in studies whose primary intent is to induce aggression or predatory behavior (Mench and Shea-Moore, 1995):

- *Isolation-induced aggression.* This involves isolating a male mouse or rat for several weeks and then staging a brief encounter (usually 5–10 minutes) with an unfamiliar group-housed male. Encounters may be staged either in the isolate's cage or in a neutral arena. If drugs are administered, they may be administered either to the isolate or to both animals. Because cues from the introduced animal can affect the outcome of the encounter, introduced mice are sometimes rendered anosmic before testing to make them less responsive to social stimulation (Stowers et al., 2002).
- *Naturalistic paradigms.* These studies aggression by placing animals in circumstances that approximate the situations that they might encounter in the wild, where they have to compete for resources, defend territories, or integrate into new social groups. Examples are introducing an 'intruder" animal into the cage or enclosure of a group of resident animals (Blanchard et al.,

1975), mixing two social groups by removing a partition between their cages (Zwirner et al., 1975), and requiring animals to compete for access to food by displacing one another from a tunnel (Miczek, 1974). Isolation of mice is not necessary to study aggression; pair-housing of a male with a female promotes consistent aggressive behavior when the male is tested in a resident-intruder situation (Fish et al., 2002). Animals may also simply be observed in their normal social groups, either in the laboratory or in the wild; this process is facilitated by the use of osmotic minipumps to deliver neuromodulators or hormones and radiotransmitters for remote collection of physiologic data.
- *Aggression modified by drugs.* Using an "intruder" paradigm, it has been shown that drugs, such as alcohol and allopregnanolone (a positive modulator of the $GABA_A$ receptor) can increase the expression of aggressive behavior in mice (Fish et al., 2002). In contrast, other drugs, such as $5\text{-}HT_{1B}$ agonists (for example, anpirtoline) will inhibit the expression of aggression (de Almeida and Miczek, 2002).
- *Predatory aggression.* This involves introducing prey species to animals, especially introducing rodents to cats and mice to rats (the muricide model). If the object of the research is to understand or influence the full predatory sequence or if the sequence ensues so rapidly after initial attack that intervention is not possible, death of the prey animal is often the endpoint. Because pain and injury to both the prey animal and predator are significant welfare issues with these kinds of studies, methods to protect the prey animal from physical attack or modeling elements of the predation sequence should be considered (Novak et al., 1998b). It may not even be necessary to use live prey. The number of times an animal serves as prey should be limited. The use of wild caught animals may be preferred due to their potentially greater experience and skill in predatory avoidance (Novak et al., 1998b). In those instances where the prey animal dies, the study should be designed to expedite the predation sequence and to minimize the pain and distress experienced by the prey animal (Huntingford, 1992).

Any situation in which unfamiliar animals are mixed or established social groups are perturbed has the potential to result in aggression, whether or not aggression is central to the aims of a study. The effects of the aggression on the recipient animal will depend on the intensity, duration, and potential for injury associated with the aggression and hence on the species being studied, the ages and sexes of the animals, and their past social experiences. If aggression is incidental to the goal of the study, many methods can be used to reduce the potential for injury, including gradual introduction of animals, allowing partial contact (for example, visual, auditory, olfactory, or tactile) before mixing and providing refuge areas to which introduced animals can escape from aggressors (Bayne and Novak, 1998). Naturalistic approaches to inducing aggression or predation may not only minimize injury but also provide information that is more reflective of the range and types of behaviors shown by animals under more ecologically relevant circumstances (Kavaliers and Choleris, 2001; Mench and Shea-Moore, 1995).

Even when aggression is a desired outcome of a study, attention should be given to minimizing injury and distress (Anonymous, 2002; Bayne and Novak, 1998; Ellwood, 1991; Huntingford, 1984). Ways of doing that include minimizing the numbers of animals used; decreasing the length of an encounter to the shortest time necessary to collect the required information, which may involve continuous observation with intervention to stop aggression at predetermined points; using artificial "model" animals rather than real animals as the recipients of aggression or the initiators of predatory encounters; placing introduced animals behind protective screens (for example, Habib et al., 2000) or barriers (Perrigo et al., 1989); and allowing the introduced or subordinate animal to control the intensity of aggression by providing refuge areas. Each of those strategies has limitations, and their usefulness will depend on the species being studied and the purpose of the study. Animals that are severely injured during an encounter should be removed as soon as possible and treated or euthanized. The use of specific animals as targets of prolonged aggression should be well justified.

## Environmental Deprivation

Animals may be exposed to nonsocial behavioral stressors to determine their effects on neural and neuroendocrine function. For example, animals may be restrained for brief or for sustained periods by being held, tethered, chaired, or immobilized by other restraint devices or placed in small enclosures or wrappings that restrict movement. Restraint may be repeated at intervals to cause intermittent stress. The animal-welfare issues associated with restraint are discussed in Chapter 3 ("Physical Restraint").

In other studies, the behavior of animals is restricted by placing them in barren environments that provide few opportunities for normal behaviors or by restricting sensory input. One or more sensory modalities (touch, audition, vision, and olfaction) may be restricted, or animals may even be kept in complete sensory isolation. The goal of such studies is generally to determine the effects of restricted environmental input on neural development. Restricted sensory or behavioral input often leads to the development of severely abnormal behaviors. Whether these effects are reversible depends on the species, the duration of restriction, and the age at which the animals are restricted. Consideration should be given to the impact of this type of research using long-lived animals due to the protracted and resilient behavior changes invoked.

## Environmental Stimulation

Stress can be induced by exposing animals to novel or extremely complex environments. The emphasis is usually on neural development, generally with a focus on fear and exploratory behaviors. Fear and exploration may be assessed with a standard battery of tests, some of which are described earlier in this chapter. Extreme novelty or complexity can have adverse physiologic and behavioral effects. However, moderate novelty and species-appropriate complexity actually have generally beneficial effects, such as enhancing neural development, learning and spatial ability, and stress competence. This is reflected in the AWRs (3.81), which mandates an appropriate plan for environmental enhancement adequate to promote the psychological well-being of nonhuman primates.

The purposeful use of environmental stimulation for experimental reasons should be distinguished from incidental, but no less stressful, stimuli that may occur in an animal facility and impact ongoing research. In either case, young animals are more susceptible to a prolonged effect of environmental stimulation and thus the use of long-lived species in this research should be well justified if the intention is to maintain the animals in the colony for extended periods of time.

## Animal Care and Use Concerns

The goal of many studies involving behavioral stressors is the induction of stress responses. Exposure to intense, repeated, or prolonged stressors can have a variety of adverse effects, including suppression of reproduction, immune dysfunction, cardiovascular and gastrointestinal impairment, and persistent disruption of neuroendocrine function (Moberg and Mench, 2000). One consequence of exposure to behavioral stressors may be the development of abnormal behaviors, including self-mutilation, mutilation of other animals (such as tail-biting in pigs and cannibalism), and stereotypic behaviors (such as bar-chewing or route-tracing). Causative factors of abnormal behaviors include social isolation, rearing in a barren environment or lack of sensory stimulation, and excessive environmental or social stimulation. Once developed, the behaviors tend to persist even when the original eliciting stimulus is removed, so the animals in question may have special husbandry and care requirements. Minimizing the duration, frequency, and intensity of stressors can minimize the effects.

When young animals are separated from their dams, parents, or broader social groups for experimental purposes, provisions must be made to care for the animals, both physically and behaviorally. In some cases, partial socialization (either temporally or physically limited contact) with peers or compatible species may be possible to mitigate the immediate stress imposed by the socially restricted environment and to improve the long-term behavioral health of the experimental animals. Alternatively, separation may be delayed until the animals are older to limit the effect of restriction. Novak et al. (1998a) suggest that young animals be monitored closely and evaluated regularly if they are separated, thus enabling more informed management decisions to address the animals' well-being.

Animal-use protocols for research involving behavioral stressors should include a thorough description of the potential animal-welfare issues associated with each stressor and a detailed plan for monitoring, record keeping, and determining when to end a test early to avoid unnecessary pain and/or distress. If little or nothing is known about the possible outcomes of exposure to a particular behavioral stressor, IACUC review and approval of the protocol may involve a requirement to conduct pilot studies, mandatory oversight of initial testing by veterinary staff, or provision of regular progress reports as a condition of continuing approval.

# References


AAALAC (Association for Assessment and Accreditation of Laboratory Animal Care International). (2002). Position Statement on Cercopthecine herpes virus 1, CHV-1 (Herpesvirus-B).

AALAS (American Association for Laboratory Animal Science). (2000). Position Statement on Recognition and Alleviation of Pain and Distress in Laboratory Animals.

Abbott, F. V., Franklin, K. B. J., and Westbrook, R. F. (1995). The formalin test: Scoring properties of the first and second phases of the pain response in rats. *Pain, 60*(1), 91–102.

Abbott, F. V., and Guy, E. R. (1995). Effects of morphine, pentobarbital and amphetamine on formalin-induced behaviours in infant rats: Sedation versus specific suppression of pain. *Pain, 62*(3), 303–12.

Abboud, T. K., Zhu, J., Richardson, M., Peres da Silva, E., and Donovan, M. (1995). Desflurane: A new volatile anesthetic for cesarean section. Maternal and neonatal effects. *Acta Anaesthesiol Scand, 39*(6), 723–26.

Abell, P., Pangilinan, G. N., and Chellman, G. J. (1995). Novel restraint device for oral dosing of rabbits. *Contemp Top Lab Anim Sci, 34*(6), 86–87.

Abo-Zena, R. A., and Horwitz, M. E. (2002). Immunomodulation in stem-cell transplantation. *Curr Opin Pharmacol, 2*(4), 452–57.

Aiello, S. (ed). (1998a). Emergency medicine and critical care. *The Merck Veterinary Manual* (8th ed). Whitehouse Station, NY: Merck and Co., Inc.

Aiello, S. (ed). (1998b). Exotic and laboratory animals. *The Merck Veterinary Manual* (8th ed). Whitehouse Station, NY: Merck and Co., Inc.

Aiello, S. (ed). (1998c). The immune system. *The Merck Veterinary Manual* (8th ed). Whitehouse Station, NY: Merck and Co., Inc.

Albee, R. R., Mattsson, J. L., Yano, B. L., and Chang, L. W. (1987). Neurobehavioral effects of dietary restriction in rats. *Neurotoxicol Teratol, 9*(3), 203–11.

Allen, A. R. (1911). Surgery of experimental lesion of spinal cord equivalent to crush injury of fracture dislocation of spinal column. *J Am Med Assoc, 57*, 878–80.

Allred, J. B., and Berntson, G. G. (1986). Is euthanasia of rats by decapitation inhumane? *J Nutr, 116*(9), 1859–61.

Amaral, D. G. (2002). The primate amygdala and the neurobiology of social behavior: Implications for understanding social anxiety. *Biol Psychiat, 51*(1), 11–17.

Anand, K. J., Coskun, V., Thrivikraman, K. V., Nemeroff, C. B., and Plotsky, P. M. (1999). Long-term behavioral effects of repetitive pain in neonatal rat pups. *Physiol Behav, 66*(4), 627–37.

Anderson, C. O., and Mason, W. A. (1974). Early experience and complexity of social organization in groups of young rhesus monkeys (Macaca mulatta). *J Comp Physiol Psychol, 87*, 681–690.

Anil, S. S., Anil, L., and Deen, J. (2002). Challenges of pain assessment in domestic animals. *J Am Vet Med Assoc, 220*(3), 313–19.

Anonymous (1966) Vitamine-A deficiency and blindness. *Lancet, 2*(7462), 536–537.

Anonymous. (1988). Anesthesia and paralysis in experimental animals: Report of a workshop held in Bethesda, Maryland, October 27, 1984. Organized by the Visual Sciences B Study Section, Division of Research Grants, National Institutes of Health. *Visual Neurosci, 1*(4), 421–26.

Anonymous (1996). Gene therapy. *Scientist, 10*(12), 14.

Anonymous (2002). Guidelines for the treatment of animals in behavioral research and teaching. *Anim Behav, 63*, 195–199.

ARENA-OLAW. (2002). *Institutional Animal Care and Use Committee Guidebook* (2nd ed). Washington, DC: US Government Printing Office.

Ator, N. A. (1991). Subjects and instrumentation. I. H. Iversen, and K. A. Lattal (eds), *Techniques in the Behavioral and Neural Sciences: Experimental Analysis of Behavior, Part 1* (pp. 1–62). Amsterdam: Elsevier Science Publishers.

Auricchio, A., Acton, P. D., Hildinger, M., Louboutin, J. P., Plossl, K., O'Connor, E., Kung, H. F., and Wilson, J. M. (2003). In vivo quantitative noninvasive imaging of gene transfer by single-photon emission computerized tomography. *Hum Gene Ther, 14*(3), 255–61.

AVMA (American Veterinary Medical Association). (2001). 2000 Report of the AVMA panel on euthanasia. *J Am Vet Med Assoc, 218*(5), 669–96.

Baker, S. N., Philbin, N., Spinks, R., Pinches, E. M., Wolpert, D. M., MacManus, D. G., Pauluis, Q., and Lemon, R. N. (1999). Multiple single unit recording in the cortex of monkeys using independently moveable microelectrodes. *J Neurosci Methods, 94*(1), 5–17.

Balaban, R. S., and Hampshire, V. A. (2001). Challenges in small animal noninvasive imaging. *ILAR J, 42*(3), 248–62.

Barlow, C., Hirotsune, S., Paylor, R., Liyanage, M., Eckhaus, M., Collins, F., Shiloh, Y., Crawley, J. N., Ried, T., Tagle, D., and Wynshaw-Boris, A. (1996). Atm-deficient mice: A paradigm of ataxia telangiectasia. *Cell, 86*(1), 159–71.

Barnes, C. A. (1979). Memory deficits associated with senescence: a neurophysiological and behavioral study in the rat. *J Comp Physiol Psychol, 93*, 74-104.

Barnes, C. A., Jung, M. W., McNaughton, B. L., Korol, D. L., Andreasson, K., and Worley, P. F. (1994). LTP saturation and spatial learning disruption: Effects of task variables and saturation levels. *J Neurosci, 14*(10), 5793–5806.

Barr, G. A. (1999). Antinociceptive effects of locally administered morphine in infant rats. *Pain, 81*(1-2), 155–61.

Barr, G. A., Miya, D. Y., and Paredes, W. (1992). Analgesic effects of intraventricular and intrathecal injection of morphine and ketocyclazocine in the infant rat. *Brain Res, 584*(1-2), 83–91.

Baumans, V., Brain, P. F., Brugere, H., Clausing, P., Jeneskog, T., Perretta, G., et al. (1994). Pain and distress in laboratory rodents and lagomorphs. *Lab Anim, 28*(2), 97–112.

Bayne, K. (1996). Normal and abnormal behaviors of laboratory animals: What do they mean? *Lab Anim, 25*, 21–24.

Bayne, K. (2000). Assessing pain and distress: A veterinary behaviorist's perspective. NRC, *Definition of Pain and Distress and Reporting Requirements for Laboratory Animals*. Washington, DC: National Academy Press.

Bayne, K. (2002). Development of the human-research animal bond and its impact on animal well-being. *ILAR J, 43*(1), 4–9.

Bayne, K., and Novak, M. (1998). Behavioral disorders. B. T. Benner, C. R. Abee, and R. Henrickson (eds), *Nonhuman Primates in Biomedical Research: Diseases*. New York: Academic Press.

Bayne, K. A. L., Beaver, B., and Mench, J. A. (2002). Laboratory animal behaviour. J. G. Fox, L. C. Anderson, F. M. Loew, and F. W. Quimby (eds), *Laboratory Animal Medicine* (2nd ed). New York: Academic Press.

Bayne, K. A. L., Dexter, S. L., and Strange, G. M. (1993). The effects of food treat provisioning and human interaction on the behavioral well-being of rhesus monkeys (*Macaca mulatta*). *Contemp Top Lab Anim Sci, 32*(2), 6–9.

Beamer, T. C. (1972). Pathological changes associated with ovarian transplantation. *44th Annual Report of the Jackson Laboratory*. Bar Harbor, ME.

Becker, H. C. (1996). Effects of ethanol on the central nervous system: Fetal damage—neurobehavioral effects. R. A. Deitrich, and V. G. Erwin (eds), *Pharmacological Effects of Ethanol on the Nervous System*. Boca Raton, FL: CRC Press.

Belcheva, M. M., Dawn, S., Barg, J., McHale, R. J., Ho, M. T., Ignatova, E., and Coscia, C. J. (1994). Transient down-regulation of neonatal rat brain mu-opioid receptors upon in utero exposure to buprenorphine. *Brain Res Dev Brain Res, 80*(1-2), 158–62.

Bellinger, L. L., and Mendel, V. E. (1975). Effect of deprivation and time of refeeding on food intake. *Physiol Behav, 14*(1), 43–46.

Belzung, C., and Griebel, G. (2001). Measuring normal and pathological anxiety-like behaviour in mice: A review. *Behav Brain Res, 125*(1-2), 141–49.

Benazzouz, A., Boraud, T., Feger, J., Burbaud, P., Bioulac, B., and Gross, C. (1996). Alleviation of experimental hemiparkinsonism by high-frequency stimulation of the subthalamic nucleus in primates: A comparison with L-dopa treatment. *Mov Disord, 11*(6), 627–32.

Benazzouz, A., Gross, C., Feger, J., Boraud, T., and Bioulac, B. (1993). Reversal of rigidity and improvement in motor performance by subthalamic high-frequency stimulation in MPTP-treated monkeys. *Eur J Neurosci, 5*(4), 382–89.

Bennett, B. T., Brown, M. J., and Schofield, J. C. (1994). *Essentials for Animal Research: A Primer for Research Personnel* (2nd ed). Beltsville, Maryland: National Agricultural Library.

Bennett, G. J., and Xie, Y. K. (1988). A peripheral mononeuropathy in rat that produces disorders of pain sensation like those seen in man. *Pain, 33*(1), 87–107.

Bergman, H., Wichmann, T., and DeLong, M. R. (1990). Reversal of experimental parkinsonism by lesions of the subthalamic nucleus. *Science, 249*(4975), 1436–38.

Bergmann, B. M., Kushida, C. A., Everson, C. A., Gilliland, M. A., Obermeyer, W., and Rechtschaffen, A. (1989). Sleep deprivation in the rat: II. Methodology. *Sleep, 12*(1), 5–12.

Bhutta, A. T., Rovnaghi, C., Simpson, P. M., Gossett, J. M., Scalzo, F. M., and Anand, K. J. (2001). Interactions of inflammatory pain and morphine in infant rats: Long- term behavioral effects. *Physiol Behav, 73*(1-2), 51–58.

Binder, R. L. (1996). Nonstressful restraint device for longitudinal evaluation and photography of mouse skin lesions during tumorigenesis studies. *Lab Anim Sci, 46*(3), 350–51.

Blanchard, R. J., Fukunaga, K., Blanchard, D. C. , and Kelley, M. J. (1975). Conspecific aggression in the laboratory rat. *J Comp Physiol Psychol, 89*(10), 1204–1209.

Blass, E. M., Cramer, C. P., and Fanselow, M. S. (1993). The development of morphine-induced antinociception in neonatal rats: A comparison of forepaw, hindpaw, and tail retraction from a thermal stimulus. *Pharmacol Biochem Behav, 44*(3), 643–49.

Blumenkopf, B., and Lipman, J. J. (1991). Studies in autotomy: Its pathophysiology and usefulness as a model of chronic pain. *Pain, 45*(2), 203–209.

Boggs, S. S. (1990). Targeted gene modification for gene therapy of stem cells. *Int J Cell Cloning, 8*(2), 80–96.

Bolles, R. C. (1970). Species-specific defensive reactions and avoidance learning. *Psychol Rev, 71*, 32–48.

Bolles, R. C. (1975). *Theory of Motivation*. New York: Harper.

Boraud, T., Bezard, E., Bioulac, B., and Gross, C. (1996). High frequency stimulation of the internal Globus Pallidus (GPi) simultaneously improves parkinsonian symptoms and reduces the firing frequency of GPi neurons in the MPTP-treated monkey. *Neurosci Lett, 215*(1), 17–20.

Bosland, M. C. (1995). Is decapitation a humane method of euthanasia in rodents? A critical review. *Contemp Top Lab Anim Sci, 34*(2), 46–48.

Bouyer, L. J., Whelan, P. J., Pearson, K. G., and Rossignol, S. (2001). Adaptive locomotor plasticity in chronic spinal cats after ankle extensors neurectomy. *J Neurosci, 21*(10), 3531–41.

Bradfield, J. F., Schachtman, T. R., McLaughlin, R. M., and Steffen, E. K. (1992). Behavioral and physiologic effects of inapparent wound infection in rats. *Lab Anim Sci, 42*(6), 572–78.

Brain, P. (1975). What does individual housing mean to a mouse? *Life Sci, 16*(2), 187–200.

Breazile, J. E. (1987). Physiologic basis and consequences of distress in animals. *J Am Vet Med Assoc, 191*(10), 1212–15.

Broadbent, J., Muccino, K. J., and Cunningham, C. L. (2002). Ethanol-induced conditioned taste aversion in 15 inbred mouse strains. *Behavioral Neuroscience, 116*(1), 138–48.

Brown, J. R., Ye, H., Bronson, R. T., Dikkes, P., and Greenberg, M. E. (1996). A defect in nurturing in mice lacking the immediate early gene fosB. *Cell, 86*(2), 297–309.

Brown, J. S., and Cunningham, C. L. (1981). The paradox of persisting self-punitive behavior. *Neurosci Biobehav Rev, 5*(3), 343–54.

Brown, M. J. (1994). Aseptic surgery for rodents. S. M. Niemi, J. S. Venable, and H. N. Guttman (eds), *Rodents and Rabbits: Current Research Issues*. Bethesda, MD: Scientists Center for Animal Welfare.

Brown, R. E., Stanford, L., and Schellinck, H. M. (2000). Developing standardized behavioral tests for knockout and mutant mice. *ILAR J, 41*(3), 163–74.

Brownlow, B. S., Park, C. R., Schwartz, R. S., and Woods, S. C. (1993). Effect of meal pattern during food restriction on body weight loss and recovery after refeeding. *Physiol Behav, 53*(3), 421–24.

Bucci, T. J. (1992). Dietary restriction: Why all the interest? An overview. *Lab Anim, 21*, 29–34.

Burnett, A. L., Calvin, D. C., Chamness, S. L., Liu, J. X., Nelson, R. J., Klein, S. L., Dawson, V. L., Dawson, T. M., and Snyder, S. H. (1997). Urinary bladder-urethral sphincter dysfunction in mice with targeted disruption of neuronal nitric oxide synthase models idiopathic voiding disorders in humans. *Nat Med, 3*(5), 571–74.

Burton, E. A., Fink, D. J., and Glorioso, J. C. (2002). Gene delivery using herpes simplex virus vectors. *DNA Cell Biol, 21*(12), 915–36.

Bush, M., Custer, R., Smeller, J., and Bush, L. M. (1977). Physiologic measures of nonhuman primates during physical restraint and chemical immobilization. *J Am Vet Med Assoc, 171*(9), 866–69.

Bushnell, P. J., Moser, V. C., MacPhail, R. C., Oshiro, W. M., Derr-Yellin, E. C., Phillips, P. M., and Kodavanti, P. R. (2002). Neurobehavioral assessments of rats perinatally exposed to a commercial mixture of polychlorinated biphenyls. *Toxicol Sci, 68*(1), 109–20.

Bushnell, P. J., Padilla, S. S., Ward, T., Pope, C. N., and Olszyk, V. B. (1991). Behavioral and neurochemical changes in rats dosed repeatedly with diisopropylfluorophosphate. *J Pharmacol Exp Ther, 256*(2), 741–50.

Cabib, S., Orsini, C., Le Moal, M., and Piazza, P. V. (2000). Abolition and reversal of strain differences in behavioral responses to drugs of abuse after a brief experience. *Science, 289*(5478), 463–65.

Carlisle, H. J., and Stock, M. J. (1993). Thermoregulatory effects of beta adrenoceptors: Effects of selective agonists and the interaction of antagonists with isoproterenol and brl-35135 in the cold. *J Pharmacol Exp Ther, 266*(3), 1446–53.

Carlstead, K., and Shepherdson, D. (2000). Alleviating stress in zoo animals with environmental enrichment. G. P. Moberg, and J. A. Mench (eds), *The Biology of Animal Stress: Basic Principles and Implications for Animal Welfare*. Wallingford, Oxon, U.K.: CAB International.

Carpenter, D. O., Hussain, R. J., Berger, D. F., Lombardo, J. P., and Park, H. Y. (2002). Electrophysiologic and behavioral effects of perinatal and acute exposure of rats to lead and polychlorinated biphenyls. *Environ Health Perspect, 110*(Suppl 3), 377–86.

Carroll, M. E., Krattiger, K. L., Gieske, D., and Sadoff, D. A. (1990). Cocaine-base smoking in rhesus monkeys: Reinforcing and physiological effects. *Psychopharmacology (Berl), 102*(4), 443–50.

Carstens, E., and Moberg, G. P. (2000). Recognizing pain and distress in laboratory animals. *ILAR J, 41*(2), 62–71.

Cases, O., Self, I., Grimsby, J., Gaspar, P., Chen, K., Pournin, S., Muller, U., Aguet, M., Babinet, C., Shih, J. C., and De Maeyer, E. (1995). Aggressive behavior and altered amounts of brain serotonin and norepinephrine in mice lacking MAOA. *Science, 268*(5218), 1763–66.

CDC-NIH (Centers for Disease Control and Prevention-National Institutes of Health). (1999). *Biosafety in Microbiological and Biomedical Laboratories* (Report No. 93-8395). Washington, DC: U.S. Government Printing Office.

Chacko, D. M., Das, A. V., Zhao, X., James, J., Bhattacharya, S., and Ahmad, I. (2003). Transplantation of ocular stem cells: The role of injury in incorporation and differentiation of grafted cells in the retina. *Vision Res, 43*(8), 937–46.

Chatham, J. C., and Blackband, S. J. (2001). Nuclear magnetic resonance spectroscopy and imaging in animal research. *ILAR J, 42*(3), 189–208.

Chen, C., Rainnie, D. G., Greene, R. W., and Tonegawa, S. (1994). Abnormal fear response and aggressive behavior in mutant mice deficient for alpha-calcium-calmodulin kinase II. *Science, 266*(5183), 291–94.

Cheng, M. Y., Bullock, C. M., Li, C., Lee, A. G., Bermak, J. C., Belluzzi, J., Weaver, D. R., Leslie, F. M., and Zhou, Q. Y. (2002). Prokineticin 2 transmits the behavioural circadian rhythm of the suprachiasmatic nucleus. *Nature, 417*(6887), 405–10.

Cherry, S. R., and Gambhir, S. S. (2001). Use of positron emission tomography in animal research. *ILAR J, 42*(3), 219–32.

Chiavegatto, S., Sun, J., Nelson, R. J., and Schnaar, R. L. (2000). A functional role for complex gangliosides: Motor deficits in GM2/GD2 synthase knockout mice. *Exp Neurol, 166*(2), 227–34.

Classen, V. (1994a). Food and water intake. J. P. Houston (ed), *Neglected Factors in Pharmacology and Neuroscience Research Techniques in the Behavioral and Neural Sciences* (Vol. 12). New York: Elsevier Science, Inc.

Classen, V. (1994b). Food restriction. J. P. Houston (ed), *Neglected Factors in Pharmacology and Neuroscience Research Techniques in the Behavioral and Neural Sciences* (Vol. 12). New York: Elsevier Science, Inc.

Clavelou, P., Dallel, R., Orliaguet, T., Woda, A., and Raboisson, P. (1995). The orofacial formalin test in rats: Effects of different formalin concentrations. *Pain, 62*(3), 295–301.

Close, B., Banister, K., Baumans, V., Bernoth, E. M., Bromage, N., Bunyan, J., Erhardt, W., Flecknell, P., Gregory, N., Hackbarth, H., Morton, D., and Warwick, C. (1996). Recommendations for euthanasia of experimental animals. *Lab Anim, 30*(4), 293–316.

Coderre, T. J., Katz, J., Vaccarino, A. L., and Melzack, R. (1993). Contribution of central neuroplasticity to pathological pain: Review of clinical and experimental evidence. *Pain, 52*(3), 259–85.

Coelho, A. M. Jr, and Carey, K. D. (1990). A social tethering system for nonhuman primates used in laboratory research. *Lab Anim Sci, 40*(4), 388–94.

Coenen, A. M., Drinkenburg, W. H., Hoenderken, R., and van Luijtelaar, E. L. (1995). Carbon dioxide euthanasia in rats: Oxygen supplementation minimizes signs of agitation and asphyxia. *Lab Anim, 29*(3), 262–68.

Cohen, H. B., and Dement, W. C. (1965). Sleep: Changes in threshold to electroconvulsive shock in rats after deprivation of "paradoxical" phase. *Science, 150*(701), 1318–19.

Cohen, J. (1988). *Statistical Power Analysis for the Behavioral Sciences* (2nd ed). Mahwah, NJ: Lawrence Erlbaum Associates.

Cohen, J. I., Davenport, D. S., Stewart, J. A., Deitchman, S., Hilliard, J. K., and Chapman, L. E. (2002). Recommendations for prevention of and therapy for exposure to B virus (cercopithecine herpesvirus 1). *Clin Infect Dis, 35*(10), 1191–1203.

Collier, G. (1989). The economics of hunger, thirst, satiety, and regulation. *Ann N Y Acad Sci, 575,* 136–56.

Collier, G., Hirsch, E., and Hamlin, P. H. (1972). The ecological determinants of reinforcement in the rat. *Physiol Behav, 9*(5), 705–16.

Collier, G., and Levitsky, D. (1967). Defense of water balance in rats: Behavioral and physiological responses to depletion. *J Comp Physiol Psychol, 64*(1), 59–67.

Colman, A. S., and Miller, J. H. (2001). Modulation of breathing by mu1 and mu2 opioid receptor stimulation in neonatal and adult rats. *Respir Physiol, 127*(2–3), 157–72.

Cook, C. J., Mellor, D. J., Harris, P. J., Ingram, J. R., and Matthews, L. R. (2000). Hands-on and hands-off measurement of stress. G. P. Moberg, and J. A. Mench (eds), *The Biology of Animal Stress: Basic Principles and Implications for Animal Welfare.* Wallingford, Oxon, U.K.: CAB International.

Cotman, C. W., and Berchtold, N. C. (2002). Exercise: A behavioral intervention to enhance brain health and plasticity. *Trends Neurosci, 25*(6), 295–301.

Crabbe, J. C., Phillips, T. J., Buck, K. J., Cunningham, C. L., and Belknap, J. K. (1999a). Identifying genes for alcohol and drug sensitivity: Recent progress and future directions. *Trends Neurosci, 22*(4), 173–79.

Crabbe, J. C., Wahlsten, D., and Dudek, B. C. (1999b). Genetics of mouse behavior: Interactions with laboratory environment. *Science, 284*(5420), 1670–72.

Crawley, J. N. (1981). Neuropharmacologic specificity of a simple animal model for the behavioral actions of benzodiazepines. *Pharmacol Biochem Behav, 15*(5), 695–99.

Crawley, J. N. (1999). Behavioral phenotyping of transgenic and knockout mice: Experimental design and evaluation of general health, sensory functions, motor abilities, and specific behavioral tests. *Brain Res, 835*(1), 18–26.

Crawley, J. N. (2000). *What's Wrong with My Mouse?* New York: Wiley.

Cristiano, R. J. (2002). Protein/DNA polyplexes for gene therapy. *Surg Oncol Clin N Am, 11*(3), 697–716.

Crofton, K. M. (1992). Reflex modification and the assessment of sensory dysfunction. H. Tilson, and C. Mitchell (eds), *Neurotoxicology* (pp. 181–211). New York: Raven Press.

Cross, H. A., and Harlow, H. F. (1965). Prolonged and progressive effects of partial isolation on the behavior of macaque monkeys. *J Exp Res Pers, 1,* 39–49.

Cryan, J. F., Page, M. E., and Lucki, I. (2002). Noradrenergic lesions differentially alter the antidepressant-like effects of reboxetine in a modified forced swim test. *Eur J Pharmacol, 436*(3), 197–205.

Cunliffe-Beamer, T. L. (1983). Biomethodology and surgical techniques. H. L. Foster, J. D. Small, and J. G. Fox (eds), *Normative Biology, Immunology and Husbandry* (Vol. III). New York: Academic Press.

Cunliffe-Beamer, T. L. (1990). Surgical techniques. H. N. Guttman (ed), *Guidelines for the Well-being of Rodents in Research.* Bethesda, MD: Scientists Center for Animal Welfare.

Cunliffe-Beamer, T. L. (1993). Applying principles of aseptic surgery to rodents. *AWIC Newsl, 4*(2), 3–6.

Cunningham, C. L., and Niehus, J. S. (1997). Flavor preference conditioning by oral self-administration of ethanol. *Psychopharmacology (Berl), 134*(3), 293–302.

Cunningham, C. L., Niehus, J. S., and Noble, D. (1993). Species difference in sensitivity to ethanol's hedonic effects. *Alcohol, 10*(2), 97–102.

Danneman, P. J., and Mandrell, T. D. (1997). Evaluation of five agents/methods for anesthesia of neonatal rats. *Lab Anim Sci, 47*(4), 386–95.

Danneman, P. J., Stein, S., and Walshaw, S. O. (1997). Humane and practical implications of using carbon dioxide mixed with oxygen for anesthesia or euthanasia of rats. *Lab Anim Sci, 47*(4), 376–85.

de Almeida, R. M., and Miczek, K. A. (2002). Aggression escalated by social instigation or by discontinuation of reinforcement ("frustration") in mice: Inhibition by anpirtoline: A 5- HT1B receptor agonist. *Neuropsychopharmacology, 27*(2), 171–81.

Delaney, S. M., and Geiger, J. D. (1996). Brain regional levels of adenosine and adenosine nucleotides in rats killed by high-energy focused microwave irradiation. *J Neurosci Methods, 64*(2), 151–56.

Dell, R. B., Holleran, S., and Ramakrishnan, R. (2002). Sample size determination. *ILAR J, 43*(4), 207–13.

Dennis, M. B. Jr. (2000). Humane endpoints for genetically engineered animal models. *ILAR J, 41*(2), 94–98.

Derr, R. F. (1991). Pain perception in decapitated rat brain. *Life Sci, 49*(19), 1399–1402.

Deyo, D. J. (1991). *Guide to Laboratory Animal Anesthesia*. Richmond, VA: Virginia Commonwealth University.

DiBartola, S. P. (2000). *Fluid Therapy in Small Animal Practice* (2nd ed). Philadelphia: W.B. Saunders.

Dilber, M. S., and Gahrton, G. (2001). Suicide gene therapy: Possible applications in haematopoietic disorders. *J Intern Med, 249*(4), 359–67.

Dohrmann, G. J., Panjabi, M. M., and Banks, D. (1978). Biomechanics of experimental spinal cord trauma. *J Neurosurg , 48*(6), 993–1001.

Domjan, M. (1998). *The Principles of Learning and Behavior* (4th ed). Pacific Grove, CA: Brooks/ Cole Publishing Company.

Donsante, A., Vogler, C., Muzyczka, N., Crawford, J. M., Barker, J., Flotte, T., Campbell-Thompson, M., Daly, T., and Sands, M. S. (2001). Observed incidence of tumorigenesis in long-term rodent studies of rAAV vectors. *Gene Ther, 8*(17), 1343–46.

Drummond, J. C., Scheller, M. S., and Todd, M. M. (1987). The effect of nitrous oxide on cortical cerebral blood flow during anesthesia with halothane and isoflurane, with and without morphine, in the rabbit. *Anesth Analg, 66*(11), 1083–89.

Dubner, R. (1987). Research on pain mechanisms in animals. *J Am Vet Med Assoc, 191*(10), 1273–76.

Dubner, R., and Ren, K. (1999). Endogenous mechanisms of sensory modulation. *Pain, Suppl 6*, S45–53.

Dubuisson, D., and Dennis, S. G. (1977). The formalin test: A quantitative study of the analgesic effects of morphine, meperidine, and brain stem stimulation in rats and cats. *Pain, 4*(2), 161–74.

Egglestone, P. A., and Wood, R. A. (1992). Management of allergies to animals. *Allergy Proc, 13*(6), 289–92.

Ellwood, R. E. (1991). Ethical implications of studies on infanticide and maternal aggression in rodents. *Anim Behav, 42*, 841–49.

Evans, H. L. (1990). Nonhuman primates in behavioral toxicology: Issues of validity, ethics and public health. *Neurotoxicol Teratol, 12*(5), 531–36.

Evans, H. L. (1994). Neurotoxicity expressed in naturally occurring behavior. B. Weiss, and J. O'Donoghe (eds), *Neurobehavioral Toxicity: Analysis and Interpretation*. New York: Raven Press.

Evans, H. L., Bushnell, P. J., Taylor, J. D., Monico, A., Teal, J. J., and Pontecorvo, M. J. (1986). A system for assessing toxicity of chemicals by continuous monitoring of homecage behaviors. *Fundam Appl Toxicol, 6*(4), 721–32.

Evans, H. L., Taylor, J. D., Ernst, J., and Graefe, J. F. (1989). Methods to evaluate the wellbeing of laboratory primates: Comparisons of macaques and tamarins. *Lab Anim Sci, 39*(4), 318–23.

Everson, C. A. (1995). Functional consequences of sustained sleep deprivation in the rat. *Behav Brain Res, 69*(1-2), 43–54.

Everson, C. A. (1997). Clinical manifestations of prolonged sleep deprivation. W. J. Schwartz (ed), *Sleep Science: Integrating Basic Research and Clinical Practice* (Vol. 15). Basel: S. Karger Publishing.

Everson, C. A., Gilliland, M. A., Kushida, C. A., Pilcher, J. J., Fang, V. S., Refetoff, S., Bergmann, B. M., and Rechtschaffen, A. (1989). Sleep deprivation in the rat: IX. Recovery. *Sleep, 12*(1), 60–67.

Everson, C. A., and Toth, L. A. (2000). Systemic bacterial invasion induced by sleep deprivation. *Am J Physiol Regul Integr Comp Physiol, 278*(4), R905–16.

Fanselow, M. S., and Cramer, C. P. (1988). The ontogeny of opiate tolerance and withdrawal in infant rats. *Pharmacol Biochem Behav, 31*(2), 431–38.

FBR (Foundation for Biomedical Research). (1987). Surgery: Protecting your animals and your study. *The Biomedical Investigator's Handbook for Researchers Using Animal Models.* Washington, DC: Foundation for Biomedical Research.

Fechter, L. D. (1995). Combined effects of noise and chemicals. *Occup Med, 10*(3), 609–21.

Fetter, M., Zee, D. S., and Proctor, L. R. (1988). Effect of lack of vision and of occipital lobectomy upon recovery from unilateral labyrinthectomy in rhesus monkey. *J Neurophysiol, 59*(2), 394–407.

Finn, D. A., Bejanian, M., Jones, B. L., Syapin, P. J., and Alkana, R. L. (1989). Temperature affects ethanol lethality in C57BL/6, 129, LS and SS mice. *Pharmacol, Biochem Behav, 34*(2), 375–80.

Fish, E. W., DeBold, J. F., and Miczek, K. A. (2002). Repeated alcohol: Behavioral sensitization and alcohol-heightened aggression in mice. *Psychopharmacology (Berl), 160*(1), 39–48.

Fitts, D. A., and St Dennis, C. (1981). Ethanol and dextrose preferences in hamsters. *J Stud Alcohol, 42*(11), 901–907.

Fitzgerald, M., and Beggs, S. (2001). The neurobiology of pain: Developmental aspects. *Neuroscientist, 7*(3), 246–57.

Fitzsimons, J. T. (1998). Angiotensin, thirst, and sodium appetite. *Physiol Rev, 78*(3), 583–686.

Flecknell, P., and Silverman, J. (2000). Pain and distress. J. Silverman, M. A. Suckow, and S. Murthy (eds), *The IACUC Handbook.* New York: CRC Press.

Flecknell, P. A. (1987). *Laboratory Animal Anesthesia: An Introduction for Research Workers and Technicians.* London: Academic Press.

Flecknell, P. A. (1996). *Laboratory Animal Anesthesia* (2nd ed). London: Academic Press.

Flecknell, P. A. (1997). Medetomidine and antipamezole: Potential uses in laboratory animals. *Lab Anim, 26*, 21–25.

Fleischman, A. R., and Chez, R. A. (1974). A chair for the chronic study of the pregnant baboon. *J Med Primatol, 3*(4), 259–64.

Fleiss, J. L. (1981). *Statistical Methods for Rates and Proportions* (2nd ed). New York: Wiley.

Fortier, L. P., Robitaille, R., and Donati, F. (2001). Increased sensitivity to depolarization and nondepolarizing neuromuscular blocking agents in young rat hemidiaphragms. *Anesthesiology, 95*(2), 478–84.

Fox, M. (1968). Veterinary ethology. M. Fox (ed), *Abnormal Behavior in Animals.* Philadelphia: W.B. Saunders Company.

Frank, M. G., Morrissette, R., and Heller, H. C. (1998). Effects of sleep deprivation in neonatal rats. *Am J Physiol, 275*(1 Pt 2), R148–57.

Fredelius, L. (1988). Time sequence of degeneration pattern of the organ of Corti after acoustic overstimulation: A transmission electron microscopy study. *Acta Otolaryngol, 106*(5–6), 373–85.

Fuchs, A. F., and Robinson, D. A. (1966). A method for measuring horizontal and vertical eye movement chronically in the monkey. *J Appl Physiol, 21*(3), 1068–70.

Fudala, P. J., and Iwamoto, E. T. (1990). Conditioned aversion after delay place conditioning with amphetamine. *Pharmacol Biochem Behav, 35*(1), 89-92.

Fujinaga, M., Doone, R., Davies, M. F., and Maze, M. (2000). Nitrous oxide lacks the antinociceptive effect on the tail flick test in newborn rats. *Anesth Analg, 91*(1), 6–10.

Galli-Taliadoros, L. A., Sedgwick, J. D., Wood, S. A., and Korner, H. (1995). Gene knock-out technology: A methodological overview for the interested novice. *J Immunol Methods, 181*(1), 1–15.

Gardiner, T. W., and Toth, L. A. (1999). Stereotactic surgery and long-term maintenance of cranial implants in research animals. *Contemp Top Lab Anim Sci, 38*(1), 56–63.

Gartner, K., Buttner, D., Dohler, K., Friedel, R., Lindena, J., and Trautschold, I. (1980). Stress response of rats to handling and experimental procedures. *Lab Anim, 14*(3), 267–74.

Gerlai, R. (1996). Gene-targeting studies of mammalian behavior: Is it the mutation or the background genotype? *Trends Neurosci, 19*(5), 177–81.

Gibbs, N. M., Larach, D. R., and Schuler, H. G. (1989). The effect of neuromuscular blockade with vecuronium on hemodynamic responses to noxious stimuli in the rat. *Anesthesiology, 71*(2), 214–17.

Gittos, M. W., and Papp, M. (2001). Antidepressant-like action of AGN 2979, a tryptophan hydroxylase activation inhibitor, in a chronic mild stress model of depression in rats. *Eur Neuropsychopharmacol, 11*(5), 351–57.

Gluck, J. P., Harlow, H. F., and Schiltz, K. A. (1973). Differential effect of early enrichment and deprivation on learning in the rhesus monkey (Macaca mulatta). *J Comp Physiol Psychol, 84*, 598–604.

Goeders, N. E., and Smith, J. E. (1987). Intracranial self-administration methodologies. *Neurosci Biobehav Rev, 11*(3), 319–29.

Goldberg, S. R., and Stolerman, I. P. (1986). *Behavioral Analysis of Drug Dependence*. Orlando: Academic Press.

Golub, M. S. (1996). Labor analgesia and infant brain development. *Pharmacol Biochem Behav, 55*(4), 619–28.

Goode, T. L., and Klein, H. J. (2002). Miniaturization: An overview of biotechnologies for monitoring the physiology and pathophysiology of rodent animal models. *Ilar J, 43*(3), 136–46.

Gordon, C. J., Becker, P., and Ali, J. S. (1998). Behavioral thermoregulatory responses of single- and group-housed mice. *Physiol Behav, 65*(2), 255–62.

Gordon, C. J., Fogelson, L., and Highfill, J. W. (1990). Hypothermia and hypometabolism: Sensitive indices of whole-body toxicity following exposure to metallic salts in the mouse. *J Toxicol Environ Health, 29*(2), 185–200.

Gordon, S. M., Brahim, J. S., Dubner, R., McCullagh, L. M., Sang, C., and Dionne, R. A. (2002). Attenuation of pain in a randomized trial by suppression of peripheral nociceptive activity in the immediate postoperative period. *Anesth Analg, 95*(5), 1351–57.

Grahame, N. J., and Cunningham, C. L. (1995). Genetic differences in intravenous cocaine self-administration between C57BL/6J and DBA/2J mice. *Psychopharmacology, 122*(3), 281–91.

Grandy, J. L., and Dunlop, C. I. (1991). Anesthesia of pups and kittens. *J Am Vet Med Assoc, 198*(7), 1244–49.

Greer, J. J., Carter, J. E., and al-Zubaidy, Z. (1995). Opioid depression of respiration in neonatal rats. *J Physiol, 485*(3), 845–55.

Griebel, G., Simiand, J., Steinberg, R., Jung, M., Gully, D., Roger, P., Geslin, M., Scatton, B., Maffrand, J. P., and Soubrie, P. (2002). 4-(2-Chloro–4-methoxy–5-methylphenyl)-N-[(1S)–2-cyclopropyl–1-(3-fluoro- 4-methylphenyl)ethyl]5-methyl-N-(2-propynyl)–1, 3-thiazol–2-amine hydrochloride (SSR-125543A), a potent and selective corticotrophin- releasing factor(1) receptor antagonist. II. Characterization in rodent models of stress-related disorders. *J Pharmacol Exp Ther, 301*(1), 333–45.

Habib, K. E., Weld, K. P., Rice, K. C., Pushkas, J., Champoux, M., Listwak, S., Webster, E. L., Atkinson, A. J., Schulkin, J., Contoreggi, C., Chrousos, G. P., McCann, S. M., Suomi, S. J., Higley, J. D., and Gold, P. W. (2000). Oral administration of a corticotropin-releasing hormone receptor antagonist significantly attenuates behavioral, neuroendocrine, and autonomic responses to stress in primates. *Proc Natl Acad Sci USA, 97*(11), 6079–84.

Hackbarth, H., Kuppers, N., and Bohnet, W. (2000). Euthanasia of rats with carbon dioxide—animal welfare aspects. *Lab Anim, 34*(1), 91–96.

Hamernik, R. P., and Qiu, W. (2001). Energy-independent factors influencing noise-induced hearing loss in the chinchilla model. *J Acoust Soc Am, 110*(6), 3163–68.

Hamilton, L. (1991). Animal Welfare Policy. *Congr Rec, 137*, E1295.

Hargreaves, K., Dubner, R., Brown, F., Flores, C., and Joris, J. (1988). A new and sensitive method for measuring thermal nociception in cutaneous hyperalgesia. *Pain, 32*(1), 77–88.

Harker, K. T., and Whishaw, I. Q. (2002). Impaired spatial performance in rats with retrosplenial lesions: Importance of the spatial problem and the rat strain in identifying lesion effects in a swimming pool. *J Neurosci, 22*(3), 1155–64.

Harlow, H. F., and Harlow, M. K. (1965). The affectional systems. A. M. Schrier, H. F. Harlow, and F. Stollnitz (eds), *Behavior of Nonhuman Primates* (Vol. 2, pp. 287–334). New York: Academic Press.

Harris, S., and Ford, S. M. (2000). Transgenic gene knock-outs: Functional genomics and therapeutic target selection. *Pharmacogenomics, 1*(4), 433–43.

Haskins, S. C., and Eisele, P. H. (1997). Postoperative support and intensive care. D. F. Kohn, S. K. Wixson, W. J. White, and G. J. Benson (eds), *Anesthesia and Analgesia in Laboratory Animals*. San Diego: Academic Press.

Hawkins, P. (2002). Recognizing and assessing pain, suffering and distress in laboratory animals: A survey of current practice in the UK with recommendations. *Lab Anim, 36*(4), 378–95.

Hedenqvist, P., and Hellebrekers, L. J. (2003). Laboratory animal analgesia, anesthesia, and euthanasia. J. Hau, and G. L. Van Hoosier (eds), *Handbook of Laboratory Animal Science: Essential Principles and Practices* (2nd ed, Vol. 1, pp. 487–520). Boca Raton: CRC Press.

Heiderstadt, K. M., McLaughlin, R. M., Wright, D. C., Walker, S. E., and Gomez-Sanchez, C. E. (2000). The effect of chronic food and water restriction on open-field behaviour and serum corticosterone levels in rats. *Lab Anim, 34*(1), 20–28.

Helmstetter, F. J., Calcagnetti, D. J., Cramer, C. P., and Fanselow, M. S. (1988). Ethylketocyclazocine and bremazocine analgesia in neonatal rats. *Pharmacol Biochem Behav, 30*(4), 817–21.

Hengemihle, J. M., Long, J. M., Betkey, J., Jucker, M., and Ingram, D. K. (1999). Age-related psychomotor and spatial learning deficits in 129/SvJ mice. *Neurobiol Aging, 20*(1), 9–18.

Hewett, T. A., Kovacs, M. S., Artwohl, J. E., and Bennett, B. T. (1993). A comparison of euthanasia methods in rats, using carbon dioxide in prefilled and fixed flow rate filled chambers. *Lab Anim Sci, 43*(6), 579–82.

Hildebrand, S. V. (1997). Paralytic agents. D. F. Kohn, S. K. Wixson, W. J. White, and G. J. Benson (eds), *Anesthesia and Analgesia in Laboratory Animals*. San Diego, CA: Academic Press.

Hillyer, E. V., and Quesenberry, K. E. (1997). *Ferrets, Rabbits and Rodents—Clinical Medicine and Surgery*. Philadelphia, PA: W.B. Saunders Co.

Hoehn, M., Nicolay, K., Franke, C., and Van der Sanden, B. (2001). Application of magnetic resonance to animal models of cerebral Ischemia. *J Magn Reson Imaging, 14*(5), 491–509.

Hofer, M. A., and Shair, H. N. (1992). Ultrasonic vocalization by rat pups during recovery from deep hypothermia. *Dev Psychobiol, 25*(7), 511–28.

Holmes, A. (2001). Targeted gene mutation approaches to the study of anxiety-like behavior in mice. *Neurosci Biobehavl Rev, 25*(3), 261–73.

Holmes, G. P., L.E. Chapman, J.A. Stewart, S.E. Straus, J.K. Hilliard, and D.S. Davenport. (1995). Guidelines for the prevention and treatment of B-virus infections in exposed persons. The B virus Working Group. *Clin. Infect. Dis., 20*, 421–39.

Holson, R. R. (1992). Euthanasia by decapitation: Evidence that this technique produces prompt, painless unconsciousness in laboratory rodents. *Neurotoxicol Teratol, 14*(4), 253–57.

Holton, L. L., Scott, E. M., Nolan, A. M., Reid, J., Welsh, E., and Flaherty, D. (1998). Comparison of three methods used for assessment of pain in dogs. *J Am Vet Med Assoc, 212*(1), 61–66.

Horne, J. A. (1985). Sleep function, with particular reference to sleep deprivation. *Ann Clin Res, 17*(5), 199–208.

Hotz, M. M., Connolly, M. S., and Lynch, C. B. (1987). Adaptation to daily meal-timing and its effect on circadian temperature rhythms in two inbred strains of mice. *Behav Genet, 17*(1), 37–51.

Huang, Y. W., Richardson, J. A., Tong, A. W., Zhang, B. Q., Stone, M. J., and Vitetta, E. S. (1993). Disseminated growth of a human multiple myeloma cell line in mice with severe combined immunodeficiency disease. *Cancer Res, 53*(6), 1392–96.

Huang, Y. W., Richardson, J. A., and Vitetta, E. S. (1995). Anti-CD54 (ICAM-1) has antitumor activity in SCID mice with human myeloma cells. *Cancer Res, 55*(3), 610–16.

Hughes, J. E., Amyx, H., Howard, J. L., Nanry, K. P., and Pollard, G. T. (1994). Health effects of water restriction to motivate lever-pressing in rats. *Lab Anim Sci, 44*(2), 135–40.

Huntingford, F. A. (1984). Some ethical issues raised by studies of predation and aggression. *Anim Behav, 32*, 210–15.

Huntingford, F. A. (1992). Some ethical issues raised by studies of predation and aggression. M. S. Dawkins, and M. Gosling (eds), *Ethics in Research in Animal Behaviour: Readings from Animal Behaviour*. London: Academic Press.

Ikarashi, Y., Sasahara, T., and Maruyama, Y. (1985). Postmortem changes in catecholamines, indoleamines, and their metabolites in rat brain regions: Prevention with 10-kW microwave irradiation. *J Neurochem, 45*(3), 935–39.

Imamura, Y., Kawamoto, H., and Nakanishi, O. (1997). Characterization of heat-hyperalgesia in an experimental trigeminal neuropathy in rats. *Exp Brain Res, 116*(1), 97–103.

Inman-Wood, S. L., Williams, M. T., Morford, L. L., and Vorhees, C. V. (2000). Effects of prenatal cocaine on Morris and Barnes maze tests of spatial learning and memory in the offspring of C57BL/6J mice. *Neurotoxicol Teratol, 22*(4), 547-57.

IRAC (Interagency Research Animal Committee). (1985). *US Government Principles for the Utilization and Care of Vertebrate Animals Used in Testing, Research, and Training*. Washington, D.C.: Office of Science and Technology Policy.

Isacson, O., Bjorklund, L. M., and Schumacher, J. M. (2003). Toward full restoration of synaptic and terminal function of the dopaminergic system in Parkinson's disease by stem cells. *Ann Neurol, 53*(3 Suppl 1), S135–48.

Iversen, I. H., and Lattal, K. A. (1991). *Experimental Analysis of Behavior, Part 1* (Techniques in the Behavioral and Neurological Sciences No. 6). Amsterdam: Elsevier.

Jaffe, J. H. (1992). Current concepts of addiction. *Res Publ Assoc Res Nerv Ment Dis, 70*, 1–21.

Jero, J., Coling, D. E., and Lalwani, A. K. (2001). The use of Preyer's reflex in evaluation of hearing in mice. *Acta Otolaryngol, 121*(5), 585–89.

Jouvet, D., Vilmont, P., Delorme, F., and Jouvet, M. (1964). Etude de la privation selective de la phase paradoxale de somneil chez le chat. *Compt. Rend. Soc. Biol., 158*, 756–59.

Jung, S. C., Han, I. P., Limaye, A., Xu, R., Gelderman, M. P., Zerfas, P., Tirumalai, K., Murray, G. J., During, M. J., Brady, R. O., and Qasba, P. (2001). Adeno-associated viral vector-mediated gene transfer results in long-term enzymatic and functional correction in multiple organs of Fabry mice. *Proc Natl Acad Sci USA, 98*(5), 2676–81.

Junghanss, C., and Marr, K. A. (2002). Infectious risks and outcomes after stem cell transplantation: Are nonmyeloablative transplants changing the picture? *Curr Opin Infect Dis, 15*(4), 347–53.

Kanal, E., Borgstede, J. P., Barkovich, A. J., Bell, C., Bradley, W. G., Felmlee, J. P., Froelich, J. W., Kaminski, E. M., Keeler, E. K., Lester, J. W., Scoumis, E. A., Zaremba, L. A., and Zinninger, M. D. (2002). American College of Radiology white paper on MR safety. *AJR Am J Roentgenol, 178*(6), 1335–47.

Kanter, G. S. (1953). Excretion and drinking after salt loading in dogs. *Am J Physiol, 174*, 87–94.

Kaplan, J. R., Manuck, S. B., Adams, M. R., Williams, J. K., Register, T. C., and Clarkson, T. B. (1993). Plaque changes and arterial enlargement in atherosclerotic monkeys after manipulation of diet and social environment. *Arterioscler Thromb, 13*(2), 254–63.

Kaplan, J. R., Manuck, S. B., Clarkson, T. B., Lusso, F. M., and Taub, D. M. (1982). Social status, environment, and atherosclerosis in cynomolgus monkeys. *Arteriosclerosis, 2*(5), 359–68.

Karaplis, A. C., Luz, A., Glowacki, J., Bronson, R. T., Tybulewicz, V. L. J., Kronenberg, H. M., and Mulligan, R. C. (1994). Lethal skeletal dysplasia from targeted disruption of the parathyroid hormone-related peptide gene. *Genes Dev 8*(3), 277–89.

Karim, A., and Arslan, M. I. (2000). Isolation modifies the behavioural response in rats. *Bangladesh Med Res Counc Bull, 26*(1), 27–32.

Kasper, S., Tauscher, J., Willeit, M., Stamenkovic, M., Neumeister, A., Kufferle, B., Barnas, C., Stastny, J., Praschak-Rieder, N., Pezawas, L., de Zwaan, M., Quiner, S., Pirker, W., Asenbaum, S., Podreka, I., and Brucke, T. (2002). Receptor and transporter imaging studies in schizophrenia, depression, bulimia and Tourette's disorder—implications for psychopharmacology. *World J Biol Psychiat, 3*(3), 133–46.

Kasten, T., Colliver, J. A., Montrey, R. D., and Dunaway, G. A. (1990). The effects of various anesthetics on tissue levels of fructose–2,6- bisphosphate in rats. *Lab Anim Sci, 40*(4), 399–401.

Kavaliers, M., and Choleris, E. (2001). Antipredator responses and defensive behavior: Ecological and ethological approaches for the neurosciences. *Neurosci Biobehav Rev, 25*(7–8), 577–86.

Kayser, V., and Guilbaud, G. (1987). Local and remote modifications of nociceptive sensitivity during carrageenin-induced inflammation in the rat. *Pain, 28*(1), 99–107.

Kelleher, R. T., and Morse, W. H. (1968). Schedules using noxious stimuli: III. Responding maintained with response produced electric shocks. *J Exp Anal Behav, 11*, 819–28.

Kerr, G. R. (1972). Nutritional requirements of subhuman primates. *Physiol Rev, 52*(2), 415–67.

Khan, T., Havey, R. M., Sayers, S. T., Patwardhan, A., and King, W. W. (1999). Animal models of spinal cord contusion injuries. *Lab Anim Sci, 49*(2), 161–72.

Kim, S. H., and Chung, J. M. (1992). An experimental model for peripheral neuropathy produced by segmental spinal nerve ligation in the rat. *Pain, 50*(3), 355–63.

King, D. L., and Arendash, G. W. (2002). Behavioral characterization of the Tg2576 transgenic model of Alzheimer's disease through 19 months. *Physiol Behav, 75*(5), 627-42.

Kirk, R. W., and Bistner, S. I. (1985). Emergency medicine and critical care: Fluid therapy. *Handbook of Veterinary Procedures and Emergency Treatment* (4th ed, pp. 591–623). Philadelphia, PA: W.B. Saunders.

Kirkpatric, L. A., and Feeney, B. C. (2000). *A Simple Guide to SPSS for Windows: Versions 8, 9, and 10* (4th ed). Belmont, CA: Wadsworth Publishing Co.

Kitchen, H., Aronson, A., Bittle, J. L., McPherson, C. W., Morton, D. B., Pakes, S. P., Rollin, B., Rowan, A. N., Sechzer, J. A., Vanderlip, J. E., Will, J. A., Clark, A. S., and Gloyd, J. S. (1987). Panel report on the colloquium on recognition and alleviation of animal pain and distress. *J Am Vet Med Assoc, 191*(10), 1186–91.

Klein, L. (1987). Neuromuscular blocking agents. C. E. Short (ed), *Principles and Practice of Veterinary Anesthesia*. Baltimore, MD: Williams and Wilkins.

Knecht, C. C. B., Allen, A. R., Williams, D. J., and Johnson, J. H. (1987). Operating room conduct. C. B. B. Knecht, D. J. Williams, A. R. Allen, and J. H. Johnson (eds), *Fundamental Techniques in Veterinary Surgery* (3rd ed). Philadelphia: W.B. Saunders Co.

Kohn, D. F., Wixson, S. K., White, W. J., and Benson, G. J. (1997). *Anesthesia and Analgesia in Laboratory Animals*. New York: Academic Press.

Koob, G. F., and Le Moal, M. (2001). Drug addiction, dysregulation of reward, and allostasis. *Neuropsychopharmacology, 24*(2), 97–129.

Koren, L., Mokady, O., Karaskov, T., Klein, J., Koren, G., and Geffen, E. (2002). A novel method using hair for determining hormonal levels in wildlife. *Anim Behav, 63*, 403–406.

Krarup, A., Chattopadhyay, P., Bhattacharjee, A. K., Burge, J. R., and Ruble, G. R. (1999). Evaluation of surrogate markers of impending death in the galactosamine- sensitized murine model of bacterial endotoxemia. *Lab Anim Sci, 49*(5), 545–50.

Kreger, M. D. (1995). *Training Materials for Animal Facility Personnel: AWIC Quick Bibliography Series, 95-08.* Beltsville, MD: National Agricultural Library.

Kritchevsky, M., and Squire, L. R. (1989). Transient global amnesia: Evidence for extensive, temporally graded retrograde amnesia. *Neurology, 39*(2 Pt 1), 213–18.

Kulig, B., Alleva, E., Bignami, G., Cohn, J., Cory-Slechta, D., Landa, V., O'Donoghue, J., and Peakall, D. (1996). Animal behavioral methods in neurotoxicity assessment: SGOMSEC joint report. *Environ Health Perspect, 104 Suppl 2*, 193–204.

Ladewig, J. (2000). Chronic intermittent stress: A model for the study of long-term stressors. G. P. Moberg, and J. A. Mench (eds), *The Biology of Animal Stress: Basic Principles and Implications for Animal Welfare*. Wallingford, Oxon, U.K.: CAB International.

Lai, C. M., Lai, Y. K., and Rakoczy, P. E. (2002). Adenovirus and adeno-associated virus vectors. *DNA Cell Biol, 21*(12), 895–913.

Lariviere, W. R., Chesler, E. J., and Mogil, J. S. (2001). Transgenic studies of pain and analgesia: Mutation or background genotype? *J Pharmacol Exp Ther, 297*(2), 467–73.

Laties, V. G. (1987). Control of animal pain and distress in behavioral studies that use food deprivation or aversive stimulation. *J Am Vet Med Assoc, 191*(10), 1290–91.

Lau, C. E., Sun, L., Wang, Q., Simpao, A., and Falk, J. L. (2000). Oral cocaine pharmacokinetics and pharmacodynamics in a cumulative-dose regimen: Pharmacokinetic-pharmacodynamic modeling of concurrent operant and spontaneous behavior within an operant context. *J Pharmacol Exp Ther, 295*(2), 634–43.

Le Belle, J. E., and Svendsen, C. N. (2002). Stem cells for neurodegenerative disorders: Where can we go from here? *BioDrugs, 16*(6), 389–401.

Le Mouellic, H., Lallemand, Y., and Brulet, P. (1990). Targeted replacement of the homeobox gene Hox–3.1 by the Escherichia coli lacZ in mouse chimeric embryos. *Proc Natl Acad Sci U S A, 87*(12), 4712–16.

Le Pen, G., and Moreau, J. L. (2002). Disruption of prepulse inhibition of startle reflex in a neurodevelopmental model of schizophrenia. Reversal by clozapine, olanzapine and risperidone but not by haloperidol. *Neuropsychopharmacology, 27*(1), 1–11.

Leach, M. C., Bowell, V. A., Allan, T. F., and Morton, D. B. (2002). Aversion to gaseous euthanasia agents in rats and mice. *Comp Med, 52*(3), 249–57.

Lein, E. S., and Shatz, C. J. (2000). Rapid regulation of brain-derived neurotrophic factor mRNA within eye- specific circuits during ocular dominance column formation. *J Neurosci, 20*(4), 1470–83.

Lemon, R. (1984a). Implantation techniques for chronic recording. *Methods for Neuronal Recording in Conscious Animals* (pp. 51–70). New York: John Wiley and Sons.

Lemon, R. (1984b). *Methods for Neuronal Recording in Conscious Animals.* Chichester: Wiley.

Lemon, R. (1984c). Preparation and training of animals for chronic recording. *Methods for Neuronal Recording in Conscious Animals* (pp. 43–49). New York: John Wiley and Sons.

Lessenich, A., Lindemann, S., Richter, A., Hedrich, H. J., Wedekind, D., Kaiser, A., and Loscher, W. (2001). A novel black-hooded mutant rat (ci3) with spontaneous circling behavior but normal auditory and vestibular functions. *Neuroscience, 107*(4), 615–28.

Levanduski, S., Bayne, K., and Dexter, S. (1992). Use of behavioral observations in the detection of diabetes mellitus. *Lab Primate Newsl, 31*(1), 14–15.

Levin, S., Semler, D., and Ruben, Z. (1993). Effects of two weeks of feed restriction on some common toxicologic parameters in Sprague-Dawley rats. *Toxicol Pathol, 21*(1), 1–14.

Lima, F. B., Hell, N. S., Timo-Iaria, C., Scivoletto, R., Dolnikoff, M. S., and Pupo, A. A. (1981). Metabolic consequences of food restriction in rats. *Physiol Behav, 27*(1), 115–23.

Line, S. W., Morgan, K. N., Markowitz, H., and Strong, S. (1989). Heart rate activity of rhesus monkeys in response to routine events. *Lab Primate Newsl, 28*(2), 9–12.

Lipman, N. S., Marini, R. P., and Flecknell, P. A. (1997). Anesthesia and analgesia in rabbits. D. F. Kohn, S. K. Wixson, W. J. White, and G. J. Benson (eds), *Anesthesia and Analgesia in Laboratory Animals*. San Diego, CA: Academic Press.

Lister, R. G. (1987). The use of a plus-maze to measure anxiety in the mouse. *Psychopharmacology, 92*(2), 180–85.

Liu, X., and Weiss, F. (2002). Reversal of ethanol-seeking behavior by D1 and D2 antagonists in an animal model of relapse: Differences in antagonist potency in previously ethanol-dependent versus nondependent rats. *J Pharmacol Exp Ther, 300*(3), 882–89.

Loeb, G. E., and Gans, C. (1986). *Electromyography for Experimentalists*. Chicago, IL: University of Chicago Press.

Lomber, S. G. (1999). The advantages and limitations of permanent or reversible deactivation techniques in the assessment of neural function. *J Neurosci Methods, 86*(2), 109–17.

Lu, Q. L., Bou-Gharios, G., and Partridge, T. A. (2003). Non-viral gene delivery in skeletal muscle: A protein factory. *Gene Ther, 10*(2), 131–42.

Luft, J., and Bode, G. (2002). Integration of safety pharmacology endpoints into toxicology studies. *Fundam Clin Pharmacol, 16*(2), 91–103.

Lukas, S. E., Griffiths, R. R., Bradford, L. D., Brady, J. V., and Daley, L. (1982). A tethering system for intravenous and intragastric drug administration in the baboon. *Pharmacol Biochem Behav, 17*(4), 823–29.

Lukas, S. E., and Moreton, J. E. (1979). A technique for chronic intragastric drug administration in the rat. *Life Sci, 25*(7), 593–600.

Luks, A. M., Zwass, M. S., Brown, R. C., Lau, M., Chari, G., and Fisher, D. M. (1998a). Opioid-induced analgesia in neonatal dogs: Pharmacodynamic differences between morphine and fentanyl. *J Pharmacol Exp Ther, 284*(1), 136–41.

Luks, F. I., Johnson, B. D., Papadakis, K., Traore, M., and Piasecki, G. J. (1998b). Predictive value of monitoring parameters in fetal surgery. *J Pediatr Surg, 33*(8), 1297–1301.

Magnusson, J. E., and Vaccarino, A. L. (1996). Reduction of autotomy following peripheral neurectomy by a single injection of bupivacaine into the cingulum bundle of rats. *Brain Res, 723*(1-2), 214–17.

Mahieu-Caputo, D., Dommergues, M., Muller, F., and Dumez, Y. (2000). Fetal pain. *Presse Med, 29*(12), 663–69.

Malmberg, A. B., Chen, C., Tonegawa, S., and Basbaum, A. I. (1997). Preserved acute pain and reduced neuropathic pain in mice lacking PKCgamma. *Science, 278*(5336), 279–83.

Manuck, S. B., Kaplan, J. R., and Clarkson, T. B. (1983a). Behaviorally induced heart rate reactivity and atherosclerosis in cynomolgus monkeys. *Psychosom Med, 45*(2), 95–108.

Manuck, S. B., Kaplan, J. R., and Clarkson, T. B. (1983b). Social instability and coronary artery atherosclerosis in cynomolgus monkeys. *Neurosci Biobehav Rev, 7*(4), 485–91.

Marsh, D., Dickenson, A., Hatch, D., and Fitzgerald, M. (1999). Epidural opioid analgesia in infant rats II: Responses to carrageenan and capsaicin. *Pain, 82*(1), 33–38.

Mason, D. E., and Brown, M. J. (1997). Monitoring of anesthesia. D. F. Kohn, S. K. Wixson, W. J. White, and G. J. Benson (eds), *Anesthesia and Analgesia in Laboratory Animals*. San Diego: Academic Press.

Mason, W. A. (1961). The effect of social restriction on the behavior of rhesus monkeys: II. Tests of gregariousness. *J Comp Physiol Psychol, 54*, 287–90.

Mason, W. A. (1968). Early social deprivation in the nohuman primates: Implication for human behavior. D. C. Glass (ed), *Environmental Influences*. New York: Rockefeller University and Russell Sage.

Mathias, R. (1996). The Basics of Brain Imaging. *NIDA Notes, 11*(5).
Mayne, M., Shepel, P. N., and Geiger, J. D. (1999). Recovery of high-integrity mRNA from brains of rats killed by high-energy focused microwave irradiation. *Brain Res Brain Res Protoc, 4*(3), 295–302.
McClearn, G. E., and Vandenbergh, D. J. (2000). Structures and limits of animal models: Examples from alcohol research. *ILAR J, 41*(3), 144–52.
McCurnin, D. M., and Jones, R. L. (1985). Principles of surgical asepsis. D. J. Slatter (ed), *Textbook of Small Animal Surgery* (Vol. 1). Philadelphia: W.B. Saunders Co.
McLaughlin, C. R., and Dewey, W. L. (1994). A comparison of the antinociceptive effects of opioid agonists in neonatal and adult rats in phasic and tonic nociceptive tests. *Pharmacol Biochem Behav, 49*(4), 1017–23.
McLaughlin, C. R., Lichtman, A. H., Fanselow, M. S., and Cramer, C. P. (1990). Tonic nociception in neonatal rats. *Pharmacol, Biochem Behav, 36*(4), 859–62.
McLay, R. N., Freeman, S. M., and Zadina, J. E. (1998). Chronic corticosterone impairs memory performance in the Barnes maze. *Physiol Behav, 63*(5), 933-7.
Mechan, A. O., O'Shea, E., Elliott, J. M., Colado, M. I., and Green, A. R. (2001). A neurotoxic dose of 3,4-methylenedioxymethamphetamine (MDMA; ecstasy) to rats results in a long-term defect in thermoregulation. *Psychopharmacology (Berl), 155*(4), 413–18.
Meisch, R. A., and Lemaire, G. A. (1993). Drug self-administration. F. van Haaren (ed), *Methods in Behavioral Pharmacology*. Amsterdam: Elsevier.
Mench, J. A. (1998). Why it is important to understand animal behavior. *ILAR J, 19*, 20–26.
Mench, J. A., and Shea-Moore, M. (1995). Moods, minds, and molecules: The neurochemistry of social behavior. *Appl Anim Behav Sci, 44*, 99–118.
Merigan, W. H., Barkdoll, E., and Maurissen, J. P. (1982). Acrylamide-induced visual impairment in primates. *Toxicol Appl Pharmacol, 62*(2), 342–45.
Mersky, H. (1979). Pain terms: A list with definitions and notes on usage. *Pain, 6*, 249–50.
Metten, P., and Crabbe, J. C. (1996). Dependence and withdrawal. R. A. Deitrich, and V. G. Erwin (eds), *Pharmacological Effects of Ethanol on the Nervous System*. Boca Raton, FL: CRC Press.
Miczek, K. A. (1974). Intraspecies aggression in rats: Effects of d-amphetamine and chlordiazepoxide. *Psychopharmacologia, 39*(4), 275–301.
Mikeska, J. A., and Klemm, W. R. (1975). EEG evaluation of humaneness of asphyxia and decapitation euthanasia of the laboratory rat. *Lab Anim Sci, 25*(2), 175–79.
Mills, C. D., Grady, J. J., and Hulsebosch, C. E. (2001). Changes in exploratory behavior as a measure of chronic central pain following spinal cord injury. *J Neurotrauma, 18*(10), 1091–1105.
Mills, C. D., Robertson, C. S., Contant, C. F., and Henley, C. M. (1997). Effects of anesthesia on polyamine metabolism and water content in the rat brain. *J Neurotrauma, 14*(12), 943–49.
Mistlberger, R., Bergmann, B., and Rechtschaffen, A. (1987). Period-amplitude analysis of rat electroencephalogram: Effects of sleep deprivation and exercise. *Sleep, 10*(6), 508–22.
Moberg, G. P. (1999). When does stress become distress. *Lab Anim, 28*, 422–26.
Moberg, G. P., and Mench, J. A. (2000). *The Biology of Animal Stress: Basic Principles and Implications for Animal Welfare*. Wallingford, Oxon, U.K.: CAB International.
Moon, P. F., Erb, H. N., Ludders, J. W., Gleed, R. D., and Pascoe, P. J. (2000). Perioperative risk factors for puppies delivered by cesarean section in the United States and Canada. *J Am Anim Hosp Assoc, 36*(4), 359–68.
Morris, R., Nosten-Bertrand, M., Dawson, T. M., Nelson, R. J., and Snyder, S. (1996). NOS and aggression. *Trends Neurosci, 19*(7), 277–78.
Morris, R. G. M. (1981). Spatial localization does not require the presence of local cues. *Learning and Motivation, 12*, 239-60.

Morrow, N. S., Schall, M., Grijalva, C. V., Geiselman, P. J., Garrick, T., Nuccion, S., and Novin, D. (1997). Body temperature and wheel running predict survival times in rats exposed to activity-stress. *Physiol Behav, 62*(4), 815–25.

Morton, D. B. (2000). A systematic approach for establishing humane endpoints. *ILAR J, 41*(2), 80–86.

Morton, D. B., and Griffiths, P. H. (1985). Guidelines on the recognition of pain, distress and discomfort in experimental animals and an hypothesis for assessment. *Vet Rec, 116*(16), 431–36.

Moser, V. C. (2000a). The functional observational battery in adult and developing rats. *Neurotoxicology, 21*(6), 989–96.

Moser, V. C. (2000b). Observational batteries in neurotoxicity testing. *Int J Toxicol, 19*, 407–11.

Mountcastle, V. B. (1980). *Medical Physiology*. St. Louis: CV Mosby Co.

Mower, G. D., and Christen, W. G. (1985). Role of visual experience in activating critical period in cat visual cortex. *J Neurophysiol, 53*(2), 572–89.

Murray, M., Kim, D., Liu, Y., Tobias, C., Tessler, A., and Fischer, I. (2002). Transplantation of genetically modified cells contributes to repair and recovery from spinal injury. *Brain Res Brain Res Rev, 40*(1–3), 292–300.

Naitoh, P., Kelly, T. L., and Englund, C. (1990). Health effects of sleep deprivation. *Occup Med, 5*(2), 209–37.

Narayanan, C. H., Fox, M. W., and Hamburger, V. (1971). Prenatal development of spontaneous and evoked activity in the rat (Rattus norvegicus albinus). *Behaviour, 40*(1), 100–134.

Nastiti, K., Benton, D., Brain, P. F., and Haug, M. (1991). The effects of 5-HT receptor ligands on ultrasonic calling in mouse pups. *Neurosci Biobehav Rev, 15*(4), 483–87.

Nelson, R. J. (1997). The use of genetic 'knockout' mice in behavioral endocrinology research. *Horm Behav, 31*(3), 188–96.

Nelson, R. J., Demas, G. E., Huang, P. L., Fishman, M. C., Dawson, V. L., Dawson, T. M., and Snyder, S. H. (1995). Behavioural abnormalities in male mice lacking neuronal nitric oxide synthase. *Nature, 378*(6555), 383–86.

Nelson, W., Cadotte, L., and Halberg, F. (1973). Circadian timing of single daily "meal" affects survival of mice. *Proc Soc Exp Biol Med, 144*(3), 766–69.

Neumann, P. E., and Collins, R. L. (1991). Genetic dissection of susceptibility to audiogenic seizures in inbred mice. *Proc Natl Acad Sci USA, 88*(12), 5408–12.

New York Academy of Sciences, and Ad Hoc Committee on Animal Research. (1988). *Interdisciplinary Principles and Guidelines for the Use of Animals in Research, Testing, and Education*. New York: New York Academy of Sciences.

Nicolaidis, S., and Rowland, N. (1975). Regulatory drinking in rats with permanent access to a bitter fluid source. *Physiol Behav, 14*(6), 819–24.

Nicolaidis, S., and Rowland, N. (1977). Intravenous self-feeding: Long-term regulation of energy balance in rats. *Science, 195*(4278), 589–91.

Niesink, R. J., van Buren-van Duinkerken, L., and van Ree, J. M. (1999). Social behavior of juvenile rats after in utero exposure to morphine: Dose-time-effect relationship. *Neuropharmacology, 38*(8), 1207–23.

NIH (National Institutes of Health). (1986). *Public Health Service Policy on Humane Care and Use of Laboratory Animals*. Washington DC: US Department of Health and Human Services.

NIH. (1991). *Preparation and Maintenance of Higher Mammals During Neuroscience Experiments: Report of a National Institutes of Health Workshop*. Bethesda, MD: NIH/National Eye Institute.

NIH. (1995). *Report and Recommendations of the Panel to Assess the NIH Investment in Research on Gene Therapy*. Washington, DC: US Government Printing Office.

NIH. (1997). *Intramural Guidelines for the Euthanasia of Mouse and Rat Fetuses and Neonates*. Available from: http://www.nal.usda.gov/awic/legislat/neonate.htm.

NIH. (1998). *Guidelines for Research Involving Recombinant DNA Molecules*. Washington, DC: US Government Printing Office.

NIH. (2002). *Methods and Welfare Considerations in Behavioral Research with Animals*. Washington, DC: U.S. Government Printing Office.

Norman, R. L., McGlone, J., and Smith, C. J. (1994). Restraint inhibits luteinizing hormone secretion in the follicular phase of the menstrual cycle in rhesus macaques. *Biol Reprod, 50*(1), 16–26.

Norman, R. L., and Smith, C. J. (1992). Restraint inhibits luteinizing hormone and testosterone secretion in intact male rhesus macaques: Effects of concurrent naloxone administration. *Neuroendocrinology, 55*(4), 405–15.

Novak, M. A., West, M., Bayne, K. A. L., and Suomi, S. J. (1998a). Ethological research methods. L. Hart (ed), *Responsible Conduct of Research in Animal Behavior* (pp. 51–66). New York: Oxford University Press.

Novak, M. A., West, M. J., Bayne, K. A., and Suomi, S. J. (1998b). Ethological research techniques and methods. L. A. Hart (ed), *Responsible Conduct with Animals in Research* (pp. 51–66). New York: Oxford University Press.

NRC (National Research Council). (1977). *Nutrient Requirements of Rabbits* (2nd ed). Washington, DC: National Academy Press.

NRC. (1985). *Nutrient Requirements of Dogs*. Washington, DC: National Academy Press.

NRC. (1986). *Nutrient Requirements of Cats*. Washington, DC: National Academy Press.

NRC. (1991). *Education and Training in the Care and Use of Laboratory Animals: A Guide for Developing Institutional Programs*. Washington, DC: National Academy Press.

NRC. (1992). *Recognition and Alleviation of Pain and Distress in Laboratory Animals*. Washington, DC: National Academy Press.

NRC. (1995). *Nutrient Requirements of Laboratory Animals* (4th ed). Washington, DC: National Academy Press.

NRC. (1996). *Guide for the Care and Use of Laboratory Animals*. Washington, DC: National Academy Press.

NRC. (1997). *Occupational Health and Safety in the Care and Use of Research Animals*. Washington, DC: National Academy Press.

NRC. (1998). *The Psychological Well-Being of Nonhuman Primates*. Washington, DC: National Academy Press.

NRC. (2000). *Definition of Pain and Distress and Reporting Requirements for Laboratory Animals*. Washington, DC: National Academy Press.

NRC. (2003a). *Occupational Health and Safety in the Care and Use of Nonhuman Primates*. Washington, DC: The National Academies Press.

NRC. (2003b). *Nutrient Requirements of Nonhuman Primates* (2nd ed). Washington, DC: The National Academies Press.

Nylander, I., Stenfors, C., Tan-No, K., Mathe, A. A., and Terenius, L. (1997). A comparison between microwave irradiation and decapitation: Basal levels of dynorphin and enkephalin and the effect of chronic morphine treatment on dynorphin peptides. *Neuropeptides, 31*(4), 357–65.

O'Farrell, V. (1996). *Manual of Canine Behaviour*. England: British Small Animal Veterinary Association.

O'Farrell, V., and Neville, P. (1994). *Manual of Feline Behaviour*. England: British Small Animal Veterinary Association.

Obernier, J. A., White, A. M., Swartzwelder, H. S., and Crews, F. T. (2002). Cognitive deficits and CNS damage after a 4-day binge ethanol exposure in rats. *Pharmacol, Biochem Behav, 72*(3), 521–32.

Olton, D. S., and Samuelson, R. J. (1976). Remembrance of places passed: spation memory in rats. *Animal Behavior Processes, 2*, 97-116.

Ostenfeld, T., and Svendsen, C. N. (2003). Recent advances in stem cell neurobiology. *Adv Tech Stand Neurosurg, 28*, 3–89.

Pare, W. P., Tejani-Butt, S., and Kluczynski, J. (2001). The emergence test: Effects of psychotropic drugs on neophobic disposition in Wistar Kyoto (WKY) and Sprague Dawley rats. *Prog Neuropsychopharmacol Biol Psychiat, 25*(8), 1615–28.

Paule, M. G. (2000). Validation of a behavioral test battery for monkeys. J. J. Buccafusco (ed), *Methods of Behavioral Analysis in Neuroscience*. Boca Raton, FL: CRC Press LLC.

Paule, M. G. (2001). Using identical behavioral tasks in children, monkeys, and rats to study the effects of drugs. *Curr Therap Res, 62*(11), 820–33.

Paule, M. G., Allen, R. R., Bailey, J. R., Scallet, A. C., Ali, S. F., Brown, R. M., and Slikker, W. Jr. (1992). Chronic marijuana smoke exposure in the rhesus monkey. II: Effects on progressive ratio and conditioned position responding. *J Pharmacol Exp Ther, 260*(1), 210–22.

Payne, A. P. (1973). A comparison of the aggressive behaviour of isolated intact and castrated male golden hamsters towards intruders introduced into the home cage. *Physiol Behav, 10*(3), 629–31.

Peck, J. W. (1978). Rats defend different body weights depending on palatability and accessibility of their food. *J Comp Physiol Psychol, 92*(3), 555–70.

Pennington, J. A. T., De Planter Bowes, A., and Nichols Church, H. (1998). *Bowes and Church's Food Values of Portions Commonly Used* (17th ed). Philadelphia: Lippincott Williams and Wilkins Publishers.

Perrigo, G., Bryant, W. C., Belvin, L., and vom Saal, F. S. (1989). The use of live pups in humane, injury-free test for infanticidal behaviour in male mice. *Anim Behav, 38*, 897–98.

Peterson, R. G. (1985). Consequences associated with nonnarcotic analgesics in the fetus and newborn. *Fed Proc, 44*(7), 2309–13.

Pfaff, D. (2001). Precision in mouse behavior genetics. *Proc Natl Acad Sci USA, 98*(11), 5957–60.

Phifer, C. B., and Terry, L. M. (1986). Use of hypothermia for general anesthesia in preweanling rodents. *Physiol Behav, 38*(6), 887–90.

Pompl, P. N., Mullan, M. J., Bjugstad, K., and Arendash, G. W. (1999). Adaptation of the circular platform spatial memory task for mice: use in detecting cognitive impairment in the APP(SW) transgenic mouse model for Alzheimer's disease. *J Neurosci Methods, 87*(1), 87-95.

Porsolt, R. D. (2000). Animal models of depression: Utility for transgenic research. *Rev Neurosci, 11*(1), 53–58.

Porsolt, R. D., Bertin, A., and Jalfre, M. (1977). Behavioral despair in mice: A primary screening test for antidepressants. *Archiv Int Pharmacodyn Ther, 229*(2), 327–36.

Porsolt, R., McArthur, R. A., and Lenegre, A. (1993). Psychotropic screening procedures. F. van Haaren (ed), *Methods in Behavioral Pharmacology*. Amsterdam: Elsevier.

Potts, W. K., Manning, C. J., and Wakeland, E. K. (1991). Mating patterns in seminatural populations of mice influenced by mhc genotype. *Nature, 352*(6336), 619–21.

Prakash, Y. S., Seckin, I., Hunter, L. W., and Sieck, G. C. (2002). Mechanisms underlying greater sensitivity of neonatal cardiac muscle to volatile anesthetics. *Anesthesiology, 96*(4), 893–906.

Quinonez, R., and Sutton, R. E. (2002). Lentiviral vectors for gene delivery into cells. *DNA Cell Biol, 21*(12), 937–51.

Ramer, M. S., French, G. D., and Bisby, M. A. (1997). Wallerian degeneration is required for both neuropathic pain and sympathetic sprouting into the DRG. *Pain, 72*(1-2), 71–78.

Rao, D. B., Moore, D. R., Reinke, L. A., and Fechter, L. D. (2001). Free radical generation in the cochlea during combined exposure to noise and carbon monoxide: An electrophysiological and an EPR study. *Hear Res, 161*(1-2), 113–22.

Rechtschaffen, A., Bergmann, B. M., Gilliland, M. A., and Bauer, K. (1999). Effects of method, duration, and sleep stage on rebounds from sleep deprivation in the rat. *Sleep, 22*(1), 11–31.

Rechtschaffen, A., Gilliland, M. A., Bergmann, B. M., and Winter, J. B. (1983). Physiological correlates of prolonged sleep deprivation in rats. *Science, 221*(4606), 182–84.

Redei, E. E., Ahmadiyeh, N., Baum, A. E., Sasso, D. A., Slone, J. L., Solberg, L. C., Will, C. C., and Volenec, A. (2001). Novel animal models of affective disorders. *Semin Clin Neuropsychiat, 6*(1), 43–67.

Redgate, E. S., Deutsch, M., and Boggs, S. S. (1991). Time of death of CNS tumor-bearing rats can be reliably predicted by body weight-loss patterns. *Lab Anim Sci, 41*(3), 269–73.

Reed, T. M., Repaske, D. R., Snyder, G. L., Greengard, P., and Vorhees, C. V. (2002). Phosphodiesterase 1B knock-out mice exhibit exaggerated locomotor hyperactivity and DARPP–32 phosphorylation in response to dopamine agonists and display impaired spatial learning. *J Neurosci, 22*(12), 5188–97.

Reicher, M. A., and Holman, E. W. (1977). Location preference and flavor aversion reinforced by amphetamine in rats. *Animal Learning & Behavior, 5*, 343-46.

Ren, K., and Dubner, R. (1993). NMDA receptor antagonists attenuate mechanical hyperalgesia in rats with unilateral inflammation of the hindpaw. *Neurosci Lett, 163*(1), 22–26.

Rescorla, R. A. (1988). Pavlovian conditioning. *Am Psychol, 43*(3), 151–60.

Richter, C. P. (1967). Sleep and activity: Their relation to the 24-hour clock. *Res Publ Assoc Res Nerv Ment Dis, 45*, 8–29.

Richter, C. P. (1971). Inborn nature of the rat's 24-hour clock. *J Comp Physiol Psychol, 75*(1), 1–4.

Ro, J. Y., Debowy, D., Lu, S., Ghosh, S., and Gardner, E. P. (1998). Digital video: A tool for correlating neuronal firing patterns with hand motor behavior. *J Neurosci Methods, 82*(2), 215–31.

Roberts, I., Kwan, I., Evans, P., and Haig, S. (2002). Does animal experimentation inform human healthcare? Observations from a systematic review of international animal experiments on fluid resuscitation. *BMJ, 324*(7335), 474–76.

Rodgers, R. J. (2001). Anxious genes, emerging themes. *Behav Pharmacol, 12*(6–7), 471–76.

Rodier, P. M., Aschner, M., Lewis, L. S., and Koeter, H. B. (1986). Cell proliferation in developing brain after brief exposure to nitrous oxide or halothane. *Anesthesiology, 64*(6), 680–87.

Rogers, T. D., Gades, N. M., Kearby, J. D., Virgous, C. K., and Dalton, J. T. (2002). Chronic restraint via tail immobilization of mice: Effects on corticosterone levels and other physiologic indices of stress. *Contemp Top Lab Anim Sci, 41*(1), 46–50.

Rolfe, P. (2000). In vivo near-infrared spectroscopy. *Annu Rev Biomed Eng, 2*, 715–54.

Rolls, B. J., and Rolls, E. T. (1981). The control of drinking. *Brit Med Bull, 37*(2), 127–30.

Rolls, B. J., Van Duijvenvoorde, P. M., and Rowe, E. A. (1983). Variety in the diet enhances intake in a meal and contributes to the development of obesity in the rat. *Physiol Behav, 31*(1), 21–27.

Rolls, B. J., Wood, R. J., Rolls, E. T., Lind, H., Lind, W., and Ledingham, J. G. (1980). Thirst following water deprivation in humans. *Am J Physiol, 239*(5), R476–82.

Rose, J. C., Macdonald, A. A., Heymann, M. A., and Rudolph, A. M. (1978). Developmental aspects of the pituitary-adrenal axis response to hemorrhagic stress in lamb fetuses in utero. *J Clin Invest, 61*(2), 42,4–32.

Rosenberg, D. P. (1991). Nonhuman primate analgesia. *Lab Anim, 20*(9), 22–32.

Ross, J. F., Mattsson, J. L., and Fix, A. S. (1998). Expanded clinical observations in toxicity studies: Historical perspectives and contemporary issues. *Regul Toxicol Pharmacol, 28*(1), 17–26.

Roth, G. S., Ingram, D. K., Black, A., and Lane, M. A. (2000). Effects of reduced energy intake on the biology of aging: The primate model. *Eur J Clin Nutr, 54*(Suppl 3), S15–S20.

Roughan, J. V., and Flecknell, P. A. (in press). Evaluation of a short duration behaviour-based post-operative pain scoring system in rats. *European Journal of Pain*.

Roughan, J. V., and Flecknell, P. A. (2000). Effects of surgery and analgesic administration on spontaneous behaviour in singly housed rats. *Res Vet Sci, 69*(3), 283–88.

Rowan, A. N. (1988). Animal anxiety and animal suffering. *Appl Anim Behav Sci, 20*, 135–42.

Rowland, N., and Flamm, C. (1977). Quinine drinking: More regulatory puzzles. *Physiol Behav, 18*(6), 1165–70.

Rowland, N. E., Morien, A., and Li, B. H. (1996). The physiology and brain mechanisms of feeding. *Nutrition, 12*(9), 626–39.

Rubinstein, M., Mogil, J. S., Japon, M., Chan, E. C., Allen, R. G., and Low, M. J. (1996). Absence of opioid stress-induced analgesia in mice lacking beta-endorphin by site-directed mutagenesis. *Proc Natl Acad Sci USA, 93*(9), 3995–4000.

Ruiz, i. Altaba A, Sanchez, P., and Dahmane, N. (2002). Gli and hedgehog in cancer: Tumours, embryos and stem cells. *Natl Rev Cancer, 2*(5), 361–72.

Rushen, J. (2000). Some issues in the interpretation of behavioral responses to stress. G. P. Moberg, and J. A. Mench (eds), *The Biology of Animal Stress: Basic Principles and Implications for Animal Welfare*. Wallingford, Oxon, U.K.: CAB International.

Russell, W. M. S., and Burch, R. L.. (1959). *The Principles of Humane Experimental Technique.* London: Methuen.

Sabik, J. F., Assad, R. S., and Hanley, F. L. (1993). Halothane as an anesthetic for fetal surgery. *J Pediatr Surg, 28*(4), 542–46; discussion 546–47.

Saleem, K. S., Pauls, J. M., Augath, M., Trinath, T., Prause, B. A., Hashikawa, T., and Logothetis, N. K. (2002). Magnetic resonance imaging of neuronal connections in the macaque monkey. *Neuron, 34*(5), 685–700.

Sanchez-Pernaute, R., Harvey-White, J., Cunningham, J., and Bankiewicz, K. S. (2001). Functional effect of adeno-associated virus mediated gene transfer of aromatic L-amino acid decarboxylase into the striatum of 6-OHDA-lesioned rats. *Mol Ther 4*(4), 324–30.

Sanford, J., Ewbank, R., Molony, V., Tavenor, W. D., and Uvarov, O. (1986). Guidelines of the recognition and assessment of pain in animals. *Vet Rec, 118*, 334–38.

Sapolsky, R. M. (1998). *Why Zebras Don't Get Ulcers*. New York: W.H. Freeman and Company.

Sarter, M., Hagan, J., and Dudchenko, P. (1992a). Behavioral screening for cognition enhancers: From indiscriminate to valid testing: Part I. *Psychopharmacology (Berl), 107*(2–3), 144–59.

Sarter, M., Hagan, J., and Dudchenko, P. (1992b). Behavioral screening for cognition enhancers: From indiscriminate to valid testing: Part II. *Psychopharmacology (Berl), 107*(4), 461–73.

SAS. (2000). *Step-By-Step Programming with Base SAS Software*. Cary, NC: SAS Institute, Inc.

Sauceda, R., and Schmidt, M. G. (2000). Refining macaque handling and restraint techniques. *Lab Anim, 29*(1), 47–49.

Saudou, F., Amara, D. A., Dierich, A., LeMeur, M., Ramboz, S., Segu, L., Buhot, M. C., and Hen, R. (1994). Enhanced aggressive behavior in mice lacking 5-HT(1B) receptor. *Science, 265*(5180), 1875–78.

Saunders, N. R., Knott, G. W., and Dziegielewska, K. M. (2000). Barriers in the immature brain. *Cell Mol Neurobiol, 20*(1), 29–40.

Savaki, H. E., Davidsen, L., Smith, C., and Sokoloff, L. (1980). Measurement of free glucose turnover in brain. *J Neurochem, 35*(2), 495–502.

Savitz, S. L., Malhotra, S., Gupta, G., and Rosenbaum, D. M. (2003). Cell transplants offer promise for stroke recovery. *J Cardiovasc Nurs, 18*(1), 57–61.

Sazani, P., Vacek, M. M., and Kole, R. (2002). Short-term and long-term modulation of gene expression by antisense therapeutics. *Curr Opin Biotechnol, 13*(5), 468–72.

Schmid, J., and Speakman, J. R. (2000). Daily energy expenditure of the grey mouse lemur (Microcebus murinus): A small primate that uses torpor. *J Comp Physiol, 170*(8), 633–41.

Schneider, J. S., and Kovelowski, C. J. 2nd. (1990). Chronic exposure to low doses of MPTP. I. Cognitive deficits in motor asymptomatic monkeys. *Brain Res, 519*(1-2), 122–28.

Schneider, M. L., Clarke, A. S., Kraemer, G. W., Roughton, E. C., Lubach, G. R., Rimm-Kaufman, S., Schmidt, D., and Ebert, M. (1998). Prenatal stress alters brain biogenic amine levels in primates. *Dev Psychopathol, 10*(3), 427–40.

Schulze, G. E., McMillan, D. E., Bailey, J. R., Scallet, A., Ali, S. F., Slikker, W. Jr, and Paule, M. G. (1988). Acute effects of delta-9-tetrahydrocannabinol in rhesus monkeys as measured by performance in a battery of complex operant tests. *J Pharmacol Exp Ther, 245*(1), 178–86.

Sedivy, J. M., and Sharp, P. A. (1989). Positive genetic selection for gene disruption in mammalian cells by homologous recombination. *Proc Natl Acad Sci USA, 86*(1), 227–31.

Seitz, P. F. D. (1959). Infantile experience and adult behaviour in animal subjects: II. Age of separation from the mother and adult behaviour in the cat. *Psychosom Med, 21*, 353–78.

Seltzer, Z., Dubner, R., and Shir, Y. (1990). A novel behavioral model of neuropathic pain disorders produced in rats by partial sciatic nerve injury. *Pain, 43*(2), 205–18.

Sharp, J., Zammit, T., Azar, T., and Lawson, D. (2002a). Does witnessing experimental procedures produce stress in male rats? *Contemp Top Lab Anim Sci, 41*(5), 8–12.

Sharp, J. L., Zammit, T. G., and Lawson, D. M. (2002b). Stress-like responses to common procedures in rats: Effect of the estrous cycle. *Contemp Top Lab Anim Sci, 41*(4), 15–22.

Sharp, J., Zammit, T., Azar, T., and Lawson, D. (2003). Are "by-stander" female Sprague-Dawley rats affected by experimental procedures? *Contemp Top Lab Anim Sci, 42*(1), 19–27.

Siegel, A., Roeling, T. A., Gregg, T. R., and Kruk, M. R. (1999). Neuropharmacology of brain-stimulation-evoked aggression. *Neurosci Biobehav Rev, 23*(3), 359–89.

Siegel, S., Hinson, R. E., Krank, M. D., and McCully, J. (1982). Heroin "overdose" death: Contribution of drug-associated environmental cues. *Science, 216*(4544), 436–37.

Silverman, J., Suckow, M., and Murthy, S. (2000). *The IACUC Handbook*. Boca Raton, FL: CRC Press LLC.

Sims, C. D., Butler, P. E., Casanova, R., Randolph, M. A., and Yaremchuk, M. J. (1997). Prolonged general anesthesia for experimental craniofacial surgery in fetal swine. *J Invest Surg, 10*(1-2), 53–57.

Singer, D. (1999). Neonatal tolerance to hypoxia: A comparative-physiological approach. *Comparative Biochemistry and Physiology—Part A: Molecular and Integrative Physiology, 123*(3), 221–34.

Smith, A. C., and Swindle, M. M. (1994). *Laboratory Animal Anesthesia, Analgesia and Surgery*. Greenbelt, MA: Scientists Center for Animal Welfare.

Smith, R. P., Gitau, R., Glover, V., and Fisk, N. M. (2000). Pain and stress in the human fetus. *Eur J Obstet Gynecol Reprod Biol, 92*(1), 161–65.

Smith, W., and Harrap, S. B. (1997). Behavioural and cardiovascular responses of rats to euthanasia using carbon dioxide gas. *Lab Anim, 31*(4), 337–46.

Snedecor, G. W., and Cochran, W. G. (1989). *Statisitical Methods* (8th ed). Ames, IA: Iowa State Press.

Soma, L. R. (1987). Assessment of animal pain in experimental animals. *Laboratory Anim Sci, 37*, 71–4.

Soothill, J. S., Morton, D. B., and Ahmad, A. (1992). The HID50 (hypothermia-inducing dose 50): An alternative to the LD50 for measurement of bacterial virulence. *Int J Exp Pathol, 73*(1), 95–98.

Soriano, P. (1995). Gene targeting in ES cells. *Annu Rev Neurosci, 18* (pp. 1–18).

Sprague, J. M. (1966). Interaction of cortex and superior colliculus in mediation of visually guided behavior in the cat. *Science, 153*(743), 1544–47.

Squire, L. R. (1986). Memory functions as affected by electroconvulsive therapy. *Ann N Y Acad Sci, 462*, 307–14.

Stafleu, F. R., Rivas, E., Rivas, T., Vorstenbosch, J., Heeger, F. R., and Beynen, A. C. (1992). The use of analogous reasoning for assessing discomfort in laboratory animals. *Anim Welfare, 1*, 77–84.

Steeghs, K., Oerlemans, F., and Wieringa, B. (1995). Mice deficient in ubiquitous mitochondrial creatine kinase are viable and fertile. *Biochim Biophys Acta—Bioenerg, 1230*(3), 130–38.

Stefanacci, L., Reber, P., Costanza, J., Wong, E., Buxton, R., Zola, S., Squire, L., and Albright, T. (1998). fMRI of monkey visual cortex. *Neuron, 20*(6), 1051–57.

Stephens, L. E., Sutherland, A. E., Klimanskaya, I. V., Andrieux, A., Meneses, J., Pedersen, R. A., and Damsky, C. H. (1995). Deletion of beta1 integrins in mice results in inner cell mass failure and peri-implantation lethality. *Genes Dev, 9*(15), 1883–95.

Stern, Y., and Langston, J. W. (1985). Intellectual changes in patients with MPTP-induced parkinsonism. *Neurology, 35*(10), 1506–1509.

Steru, L., Chermat, R., Thierry, B., and Simon, P. (1985). The tail suspension test: A new method for screening antidepressants in mice. *Psychopharmacology, 85*(3), 367–70.

Stock, H. S., Ford, K., and Wilson, M. A. (2000). Gender and gonadal hormone effects in the olfactory bulbectomy animal model of depression. *Pharmacol, Biochem Behav, 67*(1), 183–91.

Stoelting, R. K. (1999). *Pharmacology and Physiology in Anesthetic Practice* (3rd ed). Philadelphia, PA: Lippencott-Raven.

Stokes, W. S. (2000). Reducing unrelieved pain and distress in laboratory animals using humane endpoints. *ILAR J, 41*(2), 59–61.

Stowers, L., Holy, T. E., Meister, M., Dulac, C., and Koentges, G. (2002). Loss of sex discrimination and male-male aggression in mice deficient for TRP2. *Science, 295*(5559), 1493–1500.

Stricker, E. M. (1984). Biological bases of hunger and satiety: Therapeutic implications. *Nutr Rev, 42*(10), 333–40.

Stricker, E. M., and Verbalis, J. G. (1988). Hormones and behavior: The biology of thirst and sodium appetite. *Am. Sci., 76*, 261–67.

Suchecki, D., Tiba, P. A., and Tufik, S. (2002). Paradoxical sleep deprivation facilitates subsequent corticosterone response to a mild stressor in rats. *Neurosci Lett, 320*(1-2), 45–48.

Sundberg, H., Doving, K., Novikov, S., and Ursin, H. (1982). A method for studying responses and habituation to odors in rats. *Behav Neural Biol, 34*(1), 113–19.

Suomi, S. J. (1978). Maternal behavior by socially incompetent monkeys: Neglect and abuse of offspring. *J Pediatr Psychol, 3*, 28–34.

Takahashi, J. S., Pinto, L. H., and Vitaterna, M. H. (1994). Forward and reverse genetic approaches to behavior in the mouse. *Science, 264*(5166), 1724–33.

Tame, J. D., Abrams, L. M., Ding, X. Y., Yen, A., Giussani, D. A., and Nathanielsz, P. W. (1999). Level of postoperative analgesia is a critical factor in regulation of myometrial contractility after laparotomy in the pregnant baboon: Implications for human fetal surgery. *Am J Obstet Gynecol, 180*(5), 1196–1201.

Tannenbaum, J. (1999). Ethics and pain research in animals. *ILAR J, 40*(3), 97–110.

Taylor, J. D., and Evans, H. L. (1985). Effects of toluene inhalation on behavior and expired carbon dioxide in macaque monkeys. *Toxicol Appl Pharmacol, 80*(3), 487–95.

Taylor, J. R., Elsworth, J. D., Roth, R. H., Sladek, J. R. Jr, and Redmond, D. E. Jr. (1997). Severe long-term 1-methyl-4-phenyl-1,2,3,6-tetrahydropyridine-induced parkinsonism in the vervet monkey (Cercopithecus aethiops sabaeus). *Neuroscience, 81*(3), 745–55.

Taylor, W., Brown, D., Richardson, R., and Laudenslager, M. (1998). The effect of duration of individual housing on social behavior of adult male bonnet macaques (macaca radiata). *Contemp Top Lab Anim Sci, 37*(3), 47–50.

Templeton, N. S. (2002). Cationic liposome-mediated gene delivery in vivo. *Biosci Rep, 22*(2), 283–95.

Thanos, P. K., Volkow, N. D., Freimuth, P., Umegaki, H., Ikari, H., Roth, G., Ingram, D. K., and Hitzemann, R. (2001). Overexpression of dopamine D2 receptors reduces alcohol self-administration. *J Neurochem, 78*(5), 1094–1103.

Theodorsson, E., Stenfors, C., and Mathe, A. A. (1990). Microwave irradiation increases recovery of neuropeptides from brain tissues. *Peptides, 11*(6), 1191–97.

Thomas, C. E., Birkett, D., Anozie, I., Castro, M. G., and Lowenstein, P. R. (2001). Acute direct adenoviral vector cytotoxicity and chronic, but not acute, inflammatory responses correlate with decreased vector-mediated transgene expression in the brain. *Mol Ther, 3*(1), 36–46.

Threadgill, D. W., Dlugosz, A. A., Hansen, L. A., Tennenbaum, T., Lichti, U., Yee, D., LaMantia, C., Mourton, T., Herrup, K., Harris, R. C., Barnard, J. A., Yuspa, S. H., Coffey, R. J., and Magnuson, T. (1995). Targeted disruption of mouse EGF receptor: Effect of genetic background on mutant phenotype. *Science, 269*(5221), 230–34.

Thurmon, J. C., Tranquilli, W. J., and Benson, G. J. (1996). *Lumb and Jones' Veterinary Anesthesia* (3rd ed). Baltimore: Williams and Wilkins.

Todd, M. M., Weeks, J. B., and Warner, D. S. (1993). Microwave fixation for the determination of cerebral blood volume in rats. *J Cereb Blood Flow Metab, 13*(2), 328–36.

Tolwani, R. J., Jakowec, M. W., Petzinger, G. M., Green, S., and Waggie, K. (1999). Experimental models of Parkinson's disease: Insights from many models. *Lab Anim Sci, 49*(4), 363–71.

Tonegawa, S. (1994). Gene targeting: A new approach for the analysis of mammalian memory and learning. *Prog Clin Biol Res, 390*, 5–18.

Toth, L. A. (1997). The moribund state as an experimental endpoint. *Contemp Top Lab Anim Sci, 36*, 44–48.

Toth, L. A. (2000). Defining the moribund condition as an experimental endpoint for animal research. *ILAR J, 41*(2), 72–79.

Toth, L. A., and Gardiner, T. W. (1999). Stereotactic surgery and long-term maintenance of cranial implants in research animals. *Contemporary Topics (AALAS), 38*(1), 56-63.

Toth, L. A., and Gardiner, T. W. (2000). Food and water restriction protocols: Physiological and behavioral considerations. *Contemp Top Lab Anim Sci, 39*(6), 9–17.

Touzani, K., and Sclafani, A. (2002). Area postrema lesions impair flavor-toxin aversion learning but not flavor-nutrient preference learning. *Behav Neurosci, 116*(2), 256–66.

Tsukada, H., Harada, N., Nishiyama, S., Ohba, H., and Kakiuchi, T. (2000). Dose-response and duration effects of acute administrations of cocaine and GBR12909 on dopamine synthesis and transporter in the conscious monkey brain: PET studies combined with microdialysis. *Brain Res, 860*(1-2), 141–48.

Tyle, P. (1988). *Drug Delivery Devices: Fundamentals and Applications*. New York: Marcel Dekker.

UFAW (Universities Federation for Animal Welfare). (1989). Surgical procedures. *Guidelines on the Care of Laboratory Animals and Their Use for Scientific Purposes III*. London: Universities Federation for Animal Welfare.

Ullman-Cullere, M. H., and Foltz, C. J. (1999). Body condition scoring: A rapid and accurate method for assessing health status in mice. *Lab Anim Sci, 49*(3), 319–23.

Ungerstedt, U. (1968). 6-Hydroxy-dopaine induced degeneration of central monoamine neurons. *Eur J Pharmacol, 5*, 107–10.

Vainio, O., Hellsten, C., and Voipio, H. M. (2002). Pain alleviation in laboratory animals methods commonly used for perioperative pain-relief. *Scand J Lab Anim Sci, 29*(1), 1–21.

van Haaren, F. (1993). *Methods in Behavioral Pharmacology* (Techniques in Behavioral and Neurological Sciences No. 10). Amsterdam: Elsevier.

Van Leeuwen, S. D., Bonne, O. B., Avraham, Y., and Berry, E. M. (1997). Separation as a new animal model for self-induced weight loss. *Physiol Behav, 62*(1), 77–81.

Vanderwolf, C. H., Buzsaki, G., Cain, D. P., Cooley, R. K., and Robertson, B. (1988). Neocortical and hippocampal electrical activity following decapitation in the rat. *Brain Res, 451*(1-2), 340–44.

Vanhatalo, S., and van Nieuwenhuizen, O. (2000). Fetal pain? *Brain Dev, 22*(3), 145–50.

Vaugeois, J. M., Passera, G., Zuccaro, F., and Costentin, J. (1997). Individual differences in response to imipramine in the mouse tail suspension test. *Psychopharmacology, 134*(4), 387–91.

Vogler, G. A. (1997). Anesthesia equipment: Types and uses. D. F. Kohn, S. K. Wixson, W. J. White, and G. J. Benson (eds), *Anesthesia and Analgesia in Laboratory Animals*. San Diego: Academic Press.

von Borrell, E. (1995). Neuroendocrine integration of stress and significance of stress for the performance of farm animals. *Appl Anim Behav Sci, 44*, 219–28.

von Linstow Roloff, E., Platt, B., and Riedel, G. (2002). Long-term study of chronic oral aluminum exposure and spatial working memory in rats. *Behav Neurosci, 116*(2), 351–56.

Vyklicky, L. (1979). Techniques for the study of pain in animals. J. J. Bonica, J. C. Liebeskind, and D. G. Albe-Fessard (eds), *Advances in Pain Research and Therapy* (Vol. 3). New York, NY: Raven Press.

Wade, C. E., and Ortiz, R. M. (1997). Urinary excretion of cortisol from rhesus monkeys (Macaca mulatta) habituated to restraint. *Contemp Top Lab Anim Sci, 36*(5), 55–57.

Wallace, J. (2000). Humane endpoints and cancer research. *ILAR J, 41*(2), 87–93.

Wallace, J., Sanford, J., Smith, M. W., and Spencer, K. V. (1990). The assessment and control of the severity of scientific procedures on laboratory animals: Report of the laboratory animal science association working party. *Lab Anim, 24*(2), 97–130.

Warburton, D. M. (2002). Commentary on: "Comprehensive observational assessment: Ia. A systematic, quantitative procedure for assessing the behavioral and physiologic state of the mouse." *Psychopharmacology (Berl), 163*(1), 4–8.

Waszczak, B. L., Martin, L. P., Finlay, H. E., Zahr, N., and Stellar, J. R. (2002). Effects of individual and concurrent stimulation of striatal D1 and D2 dopamine receptors on electrophysiological and behavioral output from rat basal ganglia. *J Pharmacol Exp Ther, 300*(3), 850–61.

Way, E. L. (1993). Opioid tolerance and physical dependence and their relationship. A. Herz, H. Akil, and E. J. Simon (eds), *Opioids II (Handbook of Experimental Pharmacology).* Berlin: Springer-Verlag.

Wayner, M. (1964). *Thirst.* New York: The Macmillan Company.

Waynforth, H. B. (1980). *Experimental and Surgical Technique in the Rat.* London: Academic Press.

Waynforth, H. B. (1987). Standards of surgery for experimental animals. A. A. Tuffery (ed), *Laboratory Animals: An Introduction for New Experimenters.* Chichester: Wiley-Interscience.

Waynforth, H. B., Swindle, M. M., Elliott, H., and Smith, A. C. (2003). Surgery: Basic principles and procedures. J. Hau, and G. L. Van Hoosier (eds), *Handbook of Laboratory Animal Science: Essential Principles and Practices* (2nd ed, Vol. 1, pp. 487–520). Boca Raton: CRC Press.

Wechkin, S., and Breuer, L. F. (1974). Effects of isolation on aggression in the Mongolian gerbil. *Psychol Rep, 35*(1 Pt 2), 415–21.

Weisenburger, W. P. (2001). Neurotoxicology. D. Jacobson-Kram, and K. A. Keller (eds), *Toxicology Testing Handbook Principles, Applications, and Data Interpretation.* New York: Marcel Dekker, Inc.

Weiss, B., and O'Donoghue, J. L. (1994). *Neurobehavioral Toxicology.* New York: Raven Press.

Wellman, P. J., and Hoebel, B. G. (1997). *Ingestive Behavior Protocols.* New York: Society for the Study of Ingestive Behavior.

Wells, T., Windle, R. J., Peysner, K., and Forsling, M. L. (1993). Inter-colony variation in fluid balance and its relationship to vasopressin secretion in male Sprague-Dawley rats. *Lab Anim, 27*(1), 40–46.

Wenger, G. R. (1980). Cumulative dose-response curves in behavioral pharmacology. *Pharmacol Biochem Behav, 13*(5), 647–51.

Werler, M. M., Shapiro, S., and Mitchell, A. A. (1993). Periconceptional folic acid exposure and risk of occurrent neural tube defects. *J Am Med Assoc, 269*(10), 1257–61.

Wichmann, T., and DeLong, M. R. (1996). Functional and pathophysiological models of the basal ganglia. *Curr Opin Neurobiol, 6* (6), 751–58.

Williams, J. K., Kaplan, J. R., Suparto, I., Fox, J. L., and Manuck, S. B. (2003). Effects of exercise on cardiovascular outcomes in monkeys with risk factors for coronary heart disease. *Arterioscler Thromb Vasc Biol, 23*(5), 864-71.

Williams, J. K., Vita, J. A., Manuck, S. B., Selwyn, A. P., and Kaplan, J. R. (1991). Psychosocial factors impair vascular responses of coronary arteries. *Circulation, 84*(5), 2146–53.

Willner, P. (1997). Validity, reliability and utility of the chronic mild stress model of depression: A 10-year review and evaluation. *Psychopharmacology, 134*(4), 319–29.

Wixon, S. K. (1999). The role of the IACUC in assessing and managing pain and distress in research animals. V. Lukas, and M. L. Podolsky (eds), *The Care and Feeding of an IACUC: The Organization and Management of an Institutional Animal Care and Use Committee.* Boca Raton, FL: CRC Press.

Wolf, S., and Hardy, J. D. (1941). Studies on pain. Observations of pain due to loal cooling and on factors involved in the "cold pressor" effect. *J Clin Invest, 20*, 521–33.

Wolff, H. G. (1963). *Headache and Other Head Pain.* New York: Oxford University Press.

Wolfle, T. L. (1990). Policy, program and people: The three P's to well-being. J. A. Mench, and L. Krulisch (eds), *Canine Research Environment.* Bethesda: Scientists Center for Animal Welfare.

Wolfle, T. L., and Bush, R. K. (2001). The science and pervasiveness of laboratory animal allergy. *ILAR J, 42*(1), 1–3.

Wong, J. P., Saravolac, E. G., Clement, J. G., and Nagata, L. P. (1997). Development of a murine hypothermia model for study of respiratory tract influenza virus infection. *Lab Anim Sci, 47*(2), 143–47.

Wood, R. J., Rolls, E. T., and Rolls, B. J. (1982). Physiological mechanisms for thirst in the nonhuman primate. *Am J Physiol, 242*(5), R423–28.

Wrathall, J. R., Pettegrew, R. K., and Harvey, F. (1985). Spinal cord contusion in the rat: Production of graded, reproducible, injury groups. *Exp Neurol, 88*(1), 108–22.

Yang, Y., Nunes, F. A., Berencsi, K., Furth, E. E., Gonczol, E., and Wilson, J. M. (1994). Cellular immunity to viral antigens limits E1-deleted adenoviruses for gene therapy. *Proc Natl Acad Sci USA, 91*(10), 4407–11.

Yi, D. K., and Barr, G. A. (1997). Formalin-induced c-fos expression in the spinal cord of fetal rats. *Pain, 73*(3), 347–54.

Young, K. A., Berry, M. L., Mahaffey, C. L., Saionz, J. R., Hawes, N. L., Chang, B., Qing, Y. Z., Smith, R. S., Bronson, R. T., Nelson, R. J., and Simpson, E. M. (2002). Fierce: A new mouse deletion of Nr2e1; violent behaviour and ocular abnormalities are background-dependent. *Behav Brain Res, 132*(2), 145–58.

Young, L. J. (2002). The neurobiology of social recognition, approach, and avoidance. *Biol Psychiat, 51*(1), 18–26.

Yutrzenka, G. J., and Patrick, G. A. (1992). Barbiturate dependence: Neurochemcial and methodological considerations. R. R. Watson (ed), *Drugs of Abuse and Neurobiology.* Boca Raton, FL: CRC Press.

Zennou-Azogui, Y., Xerri, C., Leonard, J., and Tighilet, B. (1996). Vestibular compensation: Role of visual motion cues in the recovery of posturo-kinetic functions in the cat. *Behav Brain Res, 74*(1-2), 65–77.

Zheng, T., Steindler, D. A., and Laywell, E. D. (2002). Transplantation of an indigenous neural stem cell population leading to hyperplasia and atypical integration. *Cloning Stem Cells, 4* (1), 3–8.

Zhang, J., McQuade, J. M., Vorhees, C. V., and Xu, M. (2002). Hippocampal expression of c-fos is not essential for spatial learning. *Synapse 46*(2), 91-9.

Zhong, D. Z., Pei, C., and Xiu-Qin, L. (1996). Neurobehavioral study of prenatal exposure to hyperthermia combined with irradiation in mice. *Neurotoxicol Teratol, 18*(6), 703–709.

Zigmond, M. J., and Stricker, E. M. (1989). Animal models of parkinsonism using selective neurotoxic: Clinical and basic implications. *Int. Rev. Neurobiol, 31*, 7–79.

Zola, S. M., and Squire, L. R. (2001). Relationship between magnitude of damage to the hippocampus and impaired recognition memory in monkeys. *Hippocampus, 11*(2), 92–98.

Zwirner, P. P., Porsolt, R. D., and Loew, D. M. (1975). Inter-group aggression in mice: A new method for testing the effects of centrally active drugs. *Psychopharmacologia, 45*(2), 133–38.

# Appendix A

# Sample Size Determination

(Portions of this text are reprinted from Dell et al., 2002.)

Scientists who intend to use animals in research must justify the number of animals to be used, and committees that review proposals to use animals in research must review justification to ensure the appropriateness of the number of animals to be used. Sometimes, the number of animals to be used can be estimated best from experience; more often, a simple sample-size calculation should be performed. Even complicated experimental designs requiring sophisticated statistical models for analysis can usually be simplified to a single key or critical question so that simple formulas can be used to estimate the required sample size.

The purpose of the study may be to obtain enough tissue to analyze, to use a small number of animals for a pilot experiment, or to test a hypothesis. There is a statistical basis for estimating the number of animals (sample size) needed for several classes of hypotheses. The formula to be used depends on whether a dichotomous or continuous variable is observed and on the experimental design. Often, too few animals are used to make it possible to detect a significant effect.

## EXPERIMENTS TO TEST A FORMAL HYPOTHESIS

Most animal experiments involve formal tests of hypotheses. It is possible to estimate the number of animals required for such an experiment if a few items of information are available. Broadly, three types of variables can be measured: dichotomous variables, often expressed as rates or proportions of a yes-no outcome, such as occurrence of a disease or survival at a given time; continuous variables, such as the concentration of a substance in a body fluid or a physiologic function, such as blood flow rate or urine output; and time to occurrence of an event, such as the appearance of a disease or death. Many statistical models have been developed to test the significance of differences among means of these types of data. Detailed discussions of the models can be found in books on statistics (Cohen, 1988; Fleiss, 1981; Snedecor & Cochran, 1989), in manuals for various computer programs used for statistical analyses (Kirkpatric & Feeney, 2000; SAS, 2000), and on web sites that present elementary courses in statistics (e.g., www.ruf.rice.edu/~lane/rvls.html).

## DEFINING THE HYPOTHESIS TO BE TESTED

Although experimental designs can be complicated, an investigator's hypothesis can usually be reduced to one or a few important questions. It is then possible to compute a sample size that has a particular chance or probability of detecting (with statistical significance) an effect (or difference) that the investigator has postulated. Simple methods are presented below for computing the sample size for each of the three types of variables listed above. Note: the smaller the difference the investigator wishes to detect or the larger the population variability, the larger the sample size must be to detect a significant difference.

## EFFECT SIZE, STANDARD DEVIATION, POWER, AND SIGNIFICANCE LEVEL

In general, several factors must be known or estimated to calculate sample size: the effect size (usually the difference between two groups), the population standard deviation (for continuous data), the desired power of the experiment to detect the postulated effect, and the significance level. The first two are unique to the particular experiment; the last two are generally fixed by convention. The magnitude of the effect that the investigator wishes to detect must be stated quantitatively, and an estimate of the population standard deviation of the variable of interest must be available from a pilot study, from data obtained in a previous experiment in the investigator's laboratory, or from the scientific literature. Power is the probability of detecting a difference between treatment groups and is defined as $1-\beta$, where $\beta$ is the probability of committing a Type II error (concluding that no difference between treatment groups exists, when, in fact, there is a difference). Significance, denoted as $\alpha$, is the probability of committing a Type I error (concluding that a difference between treatment groups exists, when, in fact, there is no difference). Once values for power and significance level are chosen and the statistical model (such as chi-squared, *t*-test, analysis of variance, or linear regression) is selected, sample size can be computed by using the size of the effect that the investigator wishes to detect and the estimate of the population standard deviation of the factor to be studied.

It should be noted that in the following discussion of sample-size calculations, the aim is to simplify the question being addressed so that power calculations can be performed easily. There is no need to alter the design of the experiment and data analysis. Using, for example, randomized block, Latin square, or factorial experimental designs and the analysis of variance, it is possible to control for the effect of strain differences on such a factor as survival or response to an intervention and to obtain a more significant result than would be possible with more elementary methods. However, the simplified designs discussed here yield sample sizes close to what would be obtained with more complex analyses and therefore should help the investigator to be self-sufficient in planning experiments.

## CALCULATING SAMPLE SIZE FOR SINGLE-GROUP EXPERIMENTS

If the aim is to determine whether an event has occurred (for example, whether a pathogen is present in a colony of animals), the number of animals that need to be tested or produced is given by

$$n = \frac{\log \beta}{\log p}$$

where $\beta$ is the probability of committing a Type II error (usually 0.10 or 0.05) and $p$ represents the proportion of the animals in the colony that are not infected. Note that the proportion *not* infected is used in the formula. For example, if 30% of the animals are infected and the investigator wishes to have a 95% chance of detecting that infection, the number, $n$, of animals that are need is:

$$n = \frac{\log 0.05}{\log 0.7} = 8.4$$

Thus nine animals should be examined to have a 95% chance of detecting an infection that has affected 30% of the animals in the colony. If the prevalence of infection is lower—say, 10%—then

$$n = \frac{\log 0.05}{\log 0.9} = 28.4$$

and about 30 animals would be needed. More animals are needed if the prevalence of the pathogen is low.

## CALCULATING SAMPLE SIZE FOR CONTINUOUS VARIABLES

Experiments are often designed to measure continuous variables, such as concentration of a substance in a body fluid or blood flow rate. Although the statistical models may be complex, it is often critical to detect the difference in the mean of a variable between two groups if there is such a difference. In this case, a simple formula derived from the formula for the *t*-test can be used to compute sample size when power, significance level, size of difference in means (called the effect), and variability or standard deviation of the population means are specified:

$$n = 1 + 2C\left(\frac{s}{d}\right)^2$$

where $C$ is dependent on values chosen for significance level ($\alpha$) and power (1-$\beta$); see Table A-1. Values of $C$ for significance levels and powers not found in Table A-1 can be obtained from statistics books.

TABLE A-1 The Constant C is Dependent on the Value of $\alpha$ and 1-$\beta$

| | | $\alpha$ | |
|---|---|---|---|
| | | 0.05 | 0.01 |
| 1-$\beta$ | 0.8 | 7.85 | 11.68 |
| | 0.9 | 10.51 | 14.88 |

Suppose that a chemical that reduces appetite is to be tested to learn whether it alters the body weight of the rats. In previous experiments, the mean body weight of the rats used was 400g, with a standard deviation of 23g. Assume also that the scientist would like to be able to detect a 20g reduction in body weight between control and treated rats with a power (1-$\beta$) of 90% and a significance level ($\alpha$) of 5%. Then,

$$n = 1 + 21\left(\frac{23}{20}\right)^2 = 28.77$$

animals are needed in each group or roughly 60 animals for the whole study.

## CALCULATING SAMPLE SIZE FOR REPEAT STUDIES

Estimates of required sample size depend on the variability of the population. The greater the variability, the larger the required sample. One method of controlling for variability of a continuous variable, such as blood flow, is to measure the variable before and after an experimental intervention in a single animal, also called a paired study. In this case, instead of using an estimate of the variability of the population mean, one estimates the variability of the difference. The standard deviation of a difference in measurement in an individual is lower because it does not include interindividual variability. Stated in other terms, each animal is its own control. The number of animals needed to test a hypothesis will be reduced because the effect of animal-to-animal variation on the measurement is eliminated. Such an experiment is normally analyzed with a paired *t*-test. The following equation for *n* is derived from the paired *t*-test equation:

$$n = 2 + C\left(\frac{s}{d}\right)^2$$

Values for $C$ can be obtained from Table A-1. Note that

$$\left(\frac{s}{d}\right)^2$$

is multiplied by C in paired studies, rather than 2C, showing that a paired study is more powerful than a comparison of two independent means, as occurs in sample size calculations of continuous variables.

## SAMPLE SIZE FOR TIME TO AN EVENT

The statistical analysis of time to an event involves complicated models; however, there are two simple approaches to estimating sample size for this type of variable. The first approach is to estimate sample size by using the proportions of the experimental groups that exhibit the event by a certain time. The proportions of the experimental and control groups that exhibit an event are treated as dichotomous variables. Sample-size calculations for dichotomous variables do not require knowledge of any standard deviation. The aim of the experiment is typically to compare the proportions in two groups. If more than two groups are studied, it is often possible to identify two rates that are most important to compare.

In this method the investigator knows or can estimate the proportion of the control group that will exhibit the event and can state a difference that must be detected between the control group and the experimental group. The smaller this difference, the more animals will be needed. Thus, given estimates for proportion of the control group exhibiting the event ($p_c$) and the desired proportion of the experimental group exhibiting the event ($p_e$), then

$$n = C\frac{p_c q_c + p_e q_e}{d^2} + \frac{2}{d} + 2$$

where $q_c = 1 - p_c$; $q_e = 1 - p_e$; and $d = |p_c - p_e|$. d is the difference between $p_c$ and $p_e$, expressed as a positive quantity. Values for $C$ can be obtained from Table A-1.

Suppose that the occurrence of spontaneously developing cancer in a group of transgenic animals is 50% ($p_c = 0.5$) and the investigator wishes to test an anti-cancer drug. The investigator would like to detect when the drug causes the occurrence rate to drop to 25% of animals ($p_e = 0.25$), with a power of 90% and a significance level of 5%. Then d = .25 and $C = 10.51$ (see Table A-1 for value of C), and

$$n = 10.51 \frac{0.5 \times 0.5 + 0.25 \times 0.75}{0.25^2} + \frac{2}{0.25} + 2 = 83.57$$

animals are needed in each group, which is about 85 animals in each group, for a total number of 170 animals necessary for the experiment.

The second approach is to treat time to occurrence as a continuous variable. This approach is applicable only if all animals are followed to event occurrence (for example, until death or time to exhibit a disease, such as cancer), but it cannot be used if some animals do not reach the event during the study. To compute sample size, it is necessary to obtain the estimate of the standard deviation of the variable (s) and the magnitude of the difference (d) the investigator wishes to detect, then

$$n = 1 + 2C\left(\frac{s}{d}\right)^2$$

where C is a constant dependent on the value of $\alpha$ and $1\text{-}\beta$, as above.

Suppose that a strain of rats spontaneously develops cancer in 12 months with a standard deviation of 4 months. Assume that an investigator would like to test a drug postulated to delay the onset of cancer. If the investigator would like to be able to detect when the time to occurrence of cancer is extended to 15 months with a power of 90% and a significance level of 5%, then the difference to be detected is 3 months and 2C = 21 (C = 10.51, see Table A-1), and

$$n = 1 + 21\left(\frac{4}{3}\right)^2 = 38.37$$

animals in each group or roughly 80 animals for the whole study.

# Appendix B

# Estimating Animal Numbers

## ESTIMATING ANIMAL NUMBERS FOR BREEDING COLONIES

(Some parts of this section are reprinted from ARENA-OLAW, 2002.)

Investigators maintain breeding colonies for a variety of reasons. For example, a breeding colony may be required for an established animal model because the animal model is not commercially available, young animals with specific age or weight that cannot be provided by a commercial breeding colony are required, or the physiologic status of a mutant animal is too severely affected for it to survive shipment.

Investigators developing a new spontaneous or induced mutant animal model need to maintain their own breeding colony because there is no alternative source for the mutant. While trying to establish a breeding colony for a new mutant model, the investigator is also working to determine phenotype, to identify affected physiologic systems, and to define inheritance pattern.

Review of protocols for breeding colonies can be a challenge for the IACUC for several reasons. There may be questions about colony management, for example, the number of breeders and the number of young per cage, the breeding system (including number of females per male or continuous versus interrupted mating), the weaning age, or methods for identification of individual animals.

Large numbers of animals are required to maintain a breeding colony. The number of animals can be only approximated because it is impossible to predict the exact number and sex of offspring. There also can be confusion about whether an estimate of number of animals distinguishes between breeders, young that cannot be used in experiments because they are of the wrong genotype or sex, and animals that are actually subjected to experimental manipulations.

Determining which animals to include in the estimated number of animals in an animal-use protocol can be confusing to the investigator in the absence of IACUC-developed guidelines. The estimated number of animals that are kept for breeding purposes and not subjected to any experimental manipulations should be part of the animal-use protocol. That is in keeping with requirement to include animals "maintained but not yet used in experiments" in the USDA Annual Report of Research Facility. Each IACUC needs to develop practical guidelines about when to include young animals in the estimated number of animals. Instructions for USDA Annual Report of Research Facility do not specifically address breeding colonies except to note that animals that are used in experiments must be reported in the appropriate category.

If a study requires fertilized one-cell eggs, embryos, or fetuses, then the experimental-design section of the protocol should indicate the number of eggs, embryos, or fetuses that are required. But the estimated number of experimental animals should be limited to the number of female animals that are mated and euthanized or surgically manipulated to collect the required eggs, embryos, or fetuses. In this situation, males would be listed as breeders because they are not subjected to any experimental manipulation.

At what age to include suckling animals in the estimated number of animals is the next question. Requiring an investigator to include all animals born fails to recognize factors that result in stillbirths. Counting all live-born animals fails to recognize normal preweaning mortality. If a suckling animal will be subjected to any manipulation—such as thymectomy, toe-clipping or ear-notching for identification, tail-tip excision for genotyping, or behavioral tests—the estimated number of manipulated sucklings must be included in the number of animals used. If suckling animals will be euthanized at or before weaning because they are of the wrong genotype or sex for the experiment, they should be included as animals held but not subject to experimental manipulation.

One alternative is to instruct investigators to include all preweaning animals subjected to experimental manipulation in the estimated number of animals or for the IACUC to request estimated animal numbers as follows:

| | |
|---|---|
| Estimated number of weaned and adult animals to be subjected to experimental manipulation | ____* |
| Estimated number of suckling animals to be subjected to experimental manipulation | ____* |
| TOTAL | ____ |
| Estimated number of breeders held but not subjected to experimental manipulation | ____ |
| Estimated number of suckling animals to be euthanized at or before weaning and not subjected to experimental manipulation | ____ |

*Estimated numbers should be subdivided on the basis of invasiveness of procedures according to institutional criteria.

If a specified number of young animals are required within a 1- or 2-week period, the number of breeders required can be estimated with the mathematical formulas shown below. Mice in pair matings (1 female to 1 male) may produce more offspring than mice in trio matings (2 female to 1 male), but if the investigator needs young mice of a specific genotype or sex, trio matings may be more productive on a per-cage basis. However, there are no guarantees that the animals will breed when expected, produce the number of offspring expected, or produce animals of the expected genotypes or sex. Thus, more or fewer than the estimated number of animals may be required.

$$\text{(1)} \quad \text{Number female breeders} = \frac{\text{Number animals required for experiment}}{\text{Average pups weaned per litter} \times \text{sex correction} \times \text{mutant correction}} \times \text{infertility factor}$$

a) Sex Correction - correction when animals of a specific sex are required:

| | Multiply by: |
|---|---|
| Either sex can be used | 1 |
| Female required | 0.5* |
| Male required | 0.5* |

*Multiplier assumes 50:50 female-to-male sex ratio in offspring. If unusual sex ratio is known in advance, multiplier is modified accordingly.

b) Mutant Correction—correction when animals of a specific genotype are required:

| Inheritance pattern of mutant gene | Mating scheme* | Genotype required for experiment** | Multiply by: |
|---|---|---|---|
| Not applicable | Not applicable | No preference | 1.00 |
| Recessive | Incross | m/m | 1.00 |
| Recessive | Intercross | m/m | 0.25** |
| Recessive | Intercross | m/+ | 0.50 |
| Recessive | Backcross | m/m | 0.50*** |
| Recessive | Backcross | m/+ | 0.50 |
| Dominant | Incross | M/M | 1.00 |
| Dominant | Intercross | M/M | 0.25*** |
| Dominant | Intercross | M/+ | 0.50 |
| Dominant | Backcross | M/M | 0.50*** |
| Dominant | Backcross | M/+ | 0.50 |

*Definitions of mating schemes:

Incross Homozygous animals of the same genotype are mated together.

Cross Homozygous animals of different genotypes are mated together.

Intercross Heterozygous animals are mated together.

Backcross Homozygous animal is mated to heterozygous animal.

** Definitions of genotype abbreviations:
m/m Denotes animal homozygous for a recessive mutant gene.
m/+ Denotes animal heterozygous for a recessive mutant gene.
M/M Denotes animal homozygous for a dominant mutant gene.
M/+ Denotes animal heterozygous for a dominant mutant gene.
***This multiplier assumes homozygous mutants are fully viable. If homozygous mutants experience significant mortality, multiplier must be modified accordingly. For example, if 50% mutant animals are expected and only 75% of the homozygous mutants survive to the required age, multiplier becomes 0.38 ($0.5 \times 0.75$).

c) Infertility Factor—correction for infertility:

| Proportion infertile matings | Multiply by: |
|---|---|
| 5 % | 1.05 |
| 10 % | 1.11 |
| 15 % | 1.18 |
| 20 % | 1.25 |
| 25 % | 1.33 |
| 30 % | 1.42 |
| 35 % | 1.53 |
| 40 % | 1.66 |

(2) The number of male breeders needed will depend on the ratio of females to males. It may be 1 female to 1 male (pair matings), 2 females to 1 male (trio mating), or 3 or more females to 1 male (harem mating).

*Example:*
A study requires a group of 50 homozygous mutant female mice with a 1- or 2-week age range. Homozygous mutant mice are fully viable but sterile. The mutation is maintained by intercross mating. The mutation is maintained on inbred strain X. Strain X female breeders wean an average of 6 pups per litter. Approximately 20% of strain X matings are infertile.

$$\text{Number of female breeders required} = \frac{50}{(6 \times 0.5 \times 0.25)} \times 1.25 = 84$$

All 84 female mice are mated at the same time to synchronize litters. The number of male breeders required will vary depending on the female-to-male ratio. Depending on the strain of mice, the number of pups weaned by females in pair matings may exceed the number of pups weaned by females in trio or harem matings.

In addition to the desired 50 homozygous mutant female mice, the 84 breeders will produce on the average 50 homozygous mutant male mice, 200 heterozygous mice of both sexes, and 100 homozygous wild-type mice of both sexes. These additional 350 mice should be listed in the protocol as animals produced but not subjected to experimental manipulation.

If the study requires a sustained production over time of a specified number of pups of a specified age per week, pups weaned per female per week (or a similar productivity index) is used to estimate the number of breeder females required. The number of pups weaned per female per week is determined as follows:

$$\text{Wean per female per week} = \frac{\text{Number of pups weaned in } n \text{ weeks}}{\text{Number of female breeders}} \times \frac{1}{n \text{ weeks}}$$

Once the number of pups weaned per female per week is determined, the number of female breeders is estimated as follows:

$$\text{Number of female breeders} = \frac{\text{number of animals per week}}{\left(\text{wean per female per week} \times \text{sex correction} \times \text{mutant correction}\right)} \times \text{infertility factor} \times \frac{1}{\text{age range in weeks}}$$

*Example:*
Fifty homozygous mutant female mice are required once a month with a 2-week age range. Homozygous mutant mice are fully viable but sterile. The mutation is maintained by intercross mating on inbred strain B. Strain B female breeders wean an average of 0.6 pup, regardless of genotype, per female per week. Approximately 20% of strain B matings are infertile.

$$\text{Number of female breeders required} = \frac{50}{(0.6 \times 0.5 \times 0.25)} \times 1.25 \times \frac{1}{2} = 417$$

If productivity is to be sustained for many months, the average age of the breeding population should remain constant from month to month. That is achieved by retiring the oldest breeders and replacing them with young breeders regularly, monthly in small colonies, and weekly in very large colonies. The table below shows percentage of the colony to be replaced assuming a reproductive "life span" of 5, 6, 7, 8, 9, or 10 months. The number of replacement breeders required each month must be considered in estimating the number of female breeders required. If nonmutant-bearing animals are to be produced, the number of breeders is increased to produce the required number of replacement breeders. If mutant animals are being produced, the number of breeders may or may not need to be increased to accommodate replacement breeders, depending on the genotypes of breeders and the genotypes desired for experimental use.

| Effective reproductive life of breeders | Percentage of colony retired monthly |
|---|---|
| 5 months | 20.0 % |
| 6 months | 16.7 % |
| 7 months | 14.3 % |
| 8 months | 12.5 % |
| 9 months | 11.1 % |
| 10 months | 10.0 % |

Studies involving genetic analysis are animal-intensive. Genetic analysis can involve determining whether a single gene has dominant or recessive inheritance, identifying different genes involved in a quantitative (polygenic) trait, or fine mapping to determine chromosomal location of a mutant gene. It is possible for the investigator to estimate the number of animals required but difficult for the IACUC to evaluate the estimate in the absence of experience.

1,200 mice can be required to map a single gene with recessive inheritance and full penetrance and have adequate numbers of progeny for developmental studies, phenotyping, and linkage analysis. That number assumes a breeding colony of 10-12 pair matings with a 6- to 8-month reproductive life span, around 90% productive matings, replacement of breeders, and no unusual mutant infertility or mortality.

1,100 mice can be required for quantitative trait loci analysis using analysis of F2 progeny. That number assumes small breeding colonies of two inbred parental strains (four to six pairs) and two reciprocal F1 hybrids (two to four pairs), no unusual infertility, replacement of breeders at 6- to 8-month intervals, and generation of 500-1,000 F2 mice for genotyping.

750 mice can be required to construct a congenic strain using "speed" congenic genotyping methods. That number assumes a breeding colony of 10-12 breeding pairs, replacement of breeders, and progeny for phenotyping and genetic linkage. If the homozygous mutant does not breed and the congenic strain must be developed by using intercross matings, the estimated number of mice increases to 1,200.

After founder transgenic or targeted mice have been identified, 80-100 mice may be needed to maintain and characterize a line. The number assumes up to five breeder pairs per line, breeder replacement, no unusual infertility, and adequate numbers of weanlings for genotyping and phenotyping characterization.

## References

ARENA-OLAW. 2002. Insitutional Animal Care and Use Committee Guidebook. (2nd ed.). Washington DC: US Government Printing Office.

Festing, M.F.W. 1987. Animal production and breeding methods. In The UFAW Handbook on the Care and Management of Laboratory Animals. 6th ed. T.B. Poole (ed). Churchill Livingstone, Inc. New York. pp. 18-34.

Fox, R.R. and B.A. Witham (eds). 1997. Handbook on Genetically Standardized JX Mice. 5$^{th}$ edition. The Jackson Laboratory. Bar Harbor. Pp. 43-44, 120-125.

Standel, P.R. and Corrow, D.J. 1988. Model index of specific breeding productivity in inbred mouse colonies. Poster 48. National Meeting AALAS.

# ESTIMATING ANIMAL NUMBERS TO DEVELOP AN INDUCED MUTANT

Creation of a genetically modified mouse requires four groups of mice: stud males of the same strain or stock as the desired new model, recently weaned females of the same strain or stock as the desired new model to donate embryos for genetic modification, vasectomized males from a strain or stock with good "sex drive," and young sexually mature females from a strain or stock with good maternal characteristics. The stud males and donor females provide fertilized eggs or early embryos. Creation of a transgenic requires fertilized eggs. Creation of a targeted mutant requires blastocysts. A male mouse can be successfully mated to one or two female mice every few days to once a week, depending on the strain of mouse. A naturally ovulating inbred female mouse yields between six to eight two-cell embryos per female (Mobraaten, 1981). A naturally ovulating hybrid or outbred female mouse will usually yield more fertilized eggs or embryos. Immature female mice given hormones to induce ovulation ovulate larger numbers of eggs—16-24 eggs per inbred female (Mobraaten, 1981) to 30 or more eggs per outbred Swiss female (Wilson, 1962; Zarrow, 1961). There are marked strain differences in response to hormone injections (Hogan et al., 1986).

Generally speaking, the number of fertilized eggs collected per mouse will be higher than the number of blastocysts collected per mouse. If 100 fertilized eggs or blastocysts are to be collected, four to six female donors or 13-17 female donors could be required, depending on the genetic background of the mice, whether naturally ovulated or induced ovulated eggs are used, and whether fertilized eggs or blastocysts are collected. One report (Kovacs et al., 1993) indicated no difference in the percentage of live births between blastocysts developed from naturally ovulated donors and those developed from superovulated donors.

The number of males required depends on whether the female-to-male ratio is 1:1, 2:1, or 3:1 and higher. It is unlikely that every prospective female donor will mate at the appropriate time, conceive after mating, or respond to hormone injections. Depending on the genetic background of the mice, environmental factors, and health status, the number of unmated females or females with unfertilized eggs could be very low (less than 10%) or very high (40-50%). The number of prospective donor females must increase to compensate for females that do not mate or conceive.

Microinjection of cDNA into 100 fertilized eggs does not guarantee 100 two- or four-cell embryos to be surgically transferred. The loss will depend on technical skill, the genetic background of the mice, and environmental factors. Likewise, injection of embryonic stem cells (ES cells) into 100 blastocysts will be associated with some losses.

Vasectomized males and young sexually mature females are used to produce pseudopregnant females. Pseudopregnant female mice are required for surgical transfer of the microinjected two-cell embryos or ES cell-injected blastocysts.

The females are mated to vasectomized males. The number of vasectomized males depends on female-to-male ratios. The number of embryos surgically transferred to a pseudopregnant female is usually 8-15. If 90 microinjected two-cell embryos or blastocysts are available for surgical transfer, six to eight pseudopregnant females are required. As with collection of fertilized eggs, additional females must be mated to the vasectomized males to compensate for females that do not mate at the appropriate time.

Not all surgically transferred microinjected two-cell embryos or blastocysts undergo further cell division, implantation in the uterus, or development into viable liveborn pups. Losses depend on the genetic background of the mice, surgical skill, and environmental factors. One study reported approximately 60% births after embryo transfer of ES-cell-injected blastocysts (Kovacs et al., 1993). Blood, tail-tip, or other tissue from each liveborn pup is tested either shortly after birth or at weaning to determine whether the transgene (or targeted gene) is present in the tissues. The testing process is referred to as genotyping and is usually done with polymerase chain reaction (PCR) or Southern blot techniques. The transgene is typically present in 15-30% of the mice that develop from microinjected embryos (Gordon, 1990). Although gene alteration can be confirmed in ES cells before injection into a blastocyst, there is no guarantee that the ES cells will be incorporated into that blastocyst and produce a targeted mutant mouse.

Some young mice normally die between birth and weaning. Mortality between birth and weaning could be less than 5% or markedly higher, depending on the genetic background, induced mutation, and possible interactions between genetic background and mutation. Injection of ES cells into 100 fertilized eggs or blastocysts may not yield more than a small number of confirmed transgenic or "knock-out" weaned mice, depending on losses along the way.

Each new transgenic (or targeted) mouse must be mated to a normal mouse to determine whether the transgene (or targeted gene) is incorporated into germ cells. Each offspring is genotyped. If the first litter includes offspring that carry the transgene (or targeted gene), germline transmission has been demonstrated, and further breeding to establish the new line can began. If no offspring carrying the transgene (or targeted gene) are found, the number of offspring to be genotyped will depend on how certain one wants to be about whether the transgene (or targeted gene) is or is not incorporated into some germ cells of the parent. For example, if the new transgenic (or knockout) mouse carries the transgene in 50% of its germ cells, only half its offspring inherit the transgene (or targeted gene). Four to five offspring must be genotyped and shown not to carry the transgene (or targeted gene) before it can be concluded with 95% certainty that the new transgenic (or targeted) parent does not carry the transgene (or targeted gene) in 50% or more of its germ cells. For 99% certainty, the number of offspring increases to seven or eight. Likewise, if the transgene (or targeted gene) is present in 25% of the founder's germ cells, one-fourth of its offspring inherit the transgene (or targeted gene). Then 10 or 11 offspring must be genotyped and shown not to carry the transgene (or targeted gene) for 95% certainty. The number of offspring increases to approximately 16 for 99% certainty. If the new transgene is present in 12.5% of the founder's germ cells, one-eighth of its offspring inherit the transgene (or targeted gene). Then 22 or 23 offspring must be genotyped for 95% certainty. The number of offspring increases to approximately 34 for 99% certainty.

After a founder and first-generation offspring have been identified, mating is continued as brother x sister or backcross to a selected inbred to determine whether the transgene (or targeted gene) will transmit to later generations and can be made homozygous. As this new transgenic (or targeted) line expands from founder to first, second, and later generations, phenotypes associated with hemizygous or homozygous transgenic mice (or heterozygote or homozygote targeted mice) will become apparent. At this point, protocol review becomes identical with review for breeding colonies.

The numbers of animals required to make a new induced mutant model can be estimated by using the tables below. Any numbers generated with this table are best-guess approximations. The actual number of animals required will not be known until after germline transmission and stable inheritance have been established.

| | Number of animals |
|---|---|
| Donor females to produce desired number of naturally ovulated fertilized eggs (or blastocysts) | (i) |
| Stud males to mate with donor females | (ii) |
| Pseudopregnant females to receive manipulated embryos | (iii) |
| Vasectomized or sterile males to mate with pseudopregnant females | (ii) |
| Estimated number of nontransgenic animals to be weaned from original number of eggs (or blastocysts) | (iv) |
| Estimated number of transgenic (knockout) animals to be weaned from original number of eggs (or blastocysts) | (iv) |
| Number animals to be mated to transgenic (knockout) animals | (v) |
| Number of offspring to be genotyped | (vi) |
| Total | |

(i) N equals number of donor females needed to produce desired number of naturally ovulated fertilized eggs (blastocysts). Estimate as follows:

$$N = \frac{\text{number of blastocysts desired}}{\left(\text{number of blastocysts produced per donor female} \times \text{\% of donor females conceiving} \times \text{\% of donors females mating}\right)}$$

(ii) Will depend on whether pair (1 female to 1 male), trio (2 female to 1 male) or harem mating is used.

(iii) N equals number of pseudopregnant females to receive manipulated embryos. Estimate as follows:

$$N = \frac{\text{total number of viable embryos}}{\left(\begin{array}{c}\text{number of embryos transferred to}\\ \text{a pseudopregnant female}\end{array} \times \begin{array}{c}\text{\% of pseudopregnant females mating}\\ \text{with vasectomized males}\end{array}\right)}$$

number of viable embryos = number of blastocysts injected × expected % of blastocysts viable after injection

(iv) N equals estimated ratio of nontransgenic to transgenic animals to be weaned from original number of blastocysts. Estimate as follows:

$$N = \frac{\text{desired number of nontransgenic offspring born}}{\text{desired number of transgenic offspring born}} \times \begin{array}{c}\text{expected \% to survive}\\ \text{to breeding age}\end{array}$$

(v) N equals number of animals to be mated to transgenic animals. Estimate as follows:

N = desired number of transgenic founder × number of animals to be mated to each transgenic animal to reach breeding age

The desired number of potential founders may be predetermined by the investigator. A common number is 6-10 potential founders for a single cDNA construct.

(vi) N equals number of offspring to be genotyped. Estimate as follows:

N = number of offspring per transgenic animal × expected number of transgenic animals actually bred

## References

Beamer, W. "Use of Mutant Mice in Biological research." 12/8/89, SCAW Conference: Guidelines for the Well-being of Rodents in Research, Proceedings edited by H. Guttman.

Gordon, J.W.. 1990. Transgenic animals. Lab Animal 19(3):27-30.

Hogan, B, F Constantini and L Lacey. 1986. Manipulating the Mouse Embryo: A Laboratory Manual. Cold Springs Harbor Laboratory Press. Cold Springs Harbor, NY.

Kovacs, M.S., L Lowe and M.R. Kuehn. 1993. Use of superovulated mice as embryo donors for ES cell injected chimeras. Lab Anim Sci 43:91-93

Mobraaten, L.E.. 1981. The Jackson Laboratory Genetics Stocks Resource Repository. In Frozen Storage of Laboratory Animals. Zeilmaker, GD (ed). Gustav, Fisher, Verlag. New York. Pp. 165-1177.

Wilson E.D., and M.X. Zarrow. 1962. Comparison of superovulation in the immature mouse and rat. J Reprod. Fert. 3:148-158.

Zarrow, M.Z. and E.D. Wilson. 1961. The influence of age on superovulation in the immature rat and mouse. Endocrinology. 69:851-855.